Palgrave Studies on Norbert Elias

Series Editor
Tatiana Savoia Landini
Universidade Federal de Sao Paulo
Sao Paulo, Brazil

Despite growing, widespread appreciation for Norbert Elias's theoretical approach—often called figurational or processual sociology—there exist only a few, specialized publications on Eliasian social theory, and as of yet, no academic book series.

Palgrave Studies on Norbert Elias will therefore fill a significant gap in the market, appealing to figurationalists across disciplines: Elias's social theory is used not only in Sociology, but also Sports, Psychoanalysis/Psychology and Social Psychology, Education, Criminology, International Relations, History, Humanities (Arts, Music, and Cultural Studies), Political Science, and Public Health. Respecting the multi-disciplinary Eliasian tradition, the series is open to receiving contributions from academics outside of Sociology departments, so long as the research is grounded on Elias's approach. Publications, which shall range from Palgrave Pivots to edited collections, can be expected to explore sports, habits and manners, criminology, violence, group relations, music and musicians, theory and methods, civilizing and decivilizing processes, involvement and detachment in social sciences, formation of the modern state, power relations, and the many dozens of other topics to which Eliasian theory has been applied.

More information about this series at
http://www.palgrave.com/gp/series/16153

Artur Bogner • Stephen Mennell
Editors

Civilisations, Civilising Processes and Modernity – A Debate

Documents from the Conference at Bielefeld, 1984

between Norbert Elias, Immanuel Wallerstein, William H. McNeill, Hans-Dieter Evers, Keith Hopkins, Johann P. Arnason, Johan Goudsblom and others

palgrave
macmillan

Editors
Artur Bogner
Berlin, Germany

Stephen Mennell
School of Sociology
University College Dublin
Dublin, Ireland

ISSN 2662-3102 ISSN 2662-3110 (electronic)
Palgrave Studies on Norbert Elias
ISBN 978-3-030-80378-0 ISBN 978-3-030-80379-7 (eBook)
https://doi.org/10.1007/978-3-030-80379-7

This Palgrave Macmillan imprint is published by the registered company Springer Nature Switzerland AG.
The registered company address is: Gewerbestrasse 11, 6330 Cham, Switzerland

Acknowledgements

For their invaluable help in editing this book, we should like to thank our wives, Gabriele Rosenthal and Barbara Mennell. Professor Theresa Urbainczyk helped us with the references in Keith Hopkins's paper. We must also thank the following people and institutions for granting permission to reprint copyright material:

The Norbert Elias Foundation, Amsterdam, for permission to publish the transcript of Norbert Elias's lecture (entitled 'The formation of states and changes in restraint'), along with the transcript of the discussion, both previously unpublished.

Professor Johann P. Arnason for permission to reprint his article 'Civilisation, culture and power: reflections on Norbert Elias's genealogy of the West'. The text is identical to the second part of the revised paper that Arnason submitted to Artur Bogner in 1987. It was first published in *Thesis Eleven*, 24 (1989), pp. 44–70. The first part of the paper, which is not reproduced here, was also published in *Thesis Eleven*, 20 (1988), pp. 87–105. The original title of the paper presented at the conference was 'Large-scale units and long-term processes: Social theory and the concept of civilisation'.

Professor Hans-Dieter Evers for permission to publish his previously unpublished paper, 'The civilising process: world figuration or world system?'

Clara and Frank Goudsblom for permission to include their late father's paper, 'The domestication of fire as a civilising process', later published in revised form in *Theory, Culture and Society* 4: 2–3 (1987), pp. 457–76. For his final views on the subject, readers should consult Professor

Goudsblom's later book *Fire and Civilisation* 1992); in particular, in the light of better evidence, he repudiated the remarks he made here about the extreme duration of proto-human occupation of the Zhoukoudian 'Peking Man' site.

Professor Christopher Kelly, the late Keith Hopkins's literary executor, for permission to publish his previously unpublished paper 'From shamelessness to guilt', and his widow Dr Jennifer Hopkins, for her enthusiastic encouragement to do so.

Professor John R. McNeill for permission to reprint his late father's paper, 'The rise of the West as a long-term process', which was previously published in William H. McNeill, *Mythistory and Other Essays* (1986), pp. 43–67.

Kathy Wallerstein for permission to include her late father's paper, 'The modern world-system as a civilisation'.

Contents

Notes on Contributors

Johann P. Arnason (1940–) is Emeritus Professor of Sociology at the La Trobe University, Melbourne, Australia, and later Professor of Historical Sociology, Charles University, Prague. His published books include *The Future that Failed: Origins and Destinies of the Soviet Model* (1993), *Social Theory and Japanese Experience: The Dual Civilization* (1997), *Civilizations in Dispute: Historical Questions and Theoretical Traditions* (2003), *The Labyrinth of Modernity – Horizons, Pathways and Mutations* (2020).

Artur Bogner (1953–) is a sociologist, researcher and research fellow at the universities of Bielefeld, Essen and Bayreuth, and the Berlin-Brandenburg Academy of Sciences and Humanities (1994–1997). His publications include *Biographies in the Global South* (with co-editor Gabriele Rosenthal, 2016), *Child Soldiers in Context: Biographies, Familial and Collective Trajectories in Northern Uganda* (with co-author and co-editor G. Rosenthal, 2020), *The 1994 Civil War in Northern Ghana* (2000), *Zivilisation und Rationalisierung* (1989).

Norbert Elias (1897–1990) worked at the Zentrum für interdisziplinäre Forschung (ZiF)/Center for Interdisciplinary Research, Universität Bielefeld. He studied in Breslau, Heidelberg and Frankfurt before seeking refuge in England. Upon retiring as Reader in Sociology, University of Leicester in 1962, he served as Professor of Sociology, University of Ghana, Legon/Accra, 1962–1964. Among his renowned books are *Über den Prozess der Zivilisation* (German orig. 1939 – rev. English edn.: *On the Process of Civilisation*, 2012), *The Court Society* (German orig. 1969),

Studies on the Germans (German orig. 1989). His *Collected Works* are published at UCD Press, Dublin (18 volumes, 2006–2014).

Hans-Dieter Evers (1935–) is a development sociologist and Emeritus Professor of Development Planning and Policy, Bielefeld. His published books include *The Moral Economy of Trade: Ethnicity and Developing Markets* (1994), *Southeast Asian Urbanism* (2001), *Governing and Managing Knowledge in Asia* (2005), *Strategische Gruppen: Vergleichende Studien zu Staat, Bürokratie und Klassenbildung in der Dritten Welt* (with co-author Tilman Schiel, 1988), and *Connecting Oceans: Malaysia as a Maritime Nation* (edited with Abdul Rahman Embong and Rashila Ramli, 2020).

Johan Goudsblom (1932–2020) was Professor of Sociology at the University of Amsterdam. He was also Member of the Board of the Norbert Elias Foundation, Amsterdam. His published books in English include *Dutch Society* (1967), *Sociology in the Balance* (1977), *Nihilism and Culture* (1980), *Fire and Civilization* (1992), *Mappae Mundi: Humans and Their Habitats in Long-Term Socio-ecological Perspective* (2002).

Keith Hopkins (1934–2004) was Professor of Sociology at Brunel University, London, and subsequently Professor of Ancient History, University of Cambridge (1985–2000). His published books include *Conquerors and Slaves* (1978), *Death and Renewal* (1983), and *A World Full of Gods* (1999).

William H. McNeill (1917–2016) was Professor of History at the University of Chicago. His published books include *The Rise of the West* (1963), *Europe's Steppe Frontier, 1500–1800* (1964), *A World History* (1967), *Venice: The Hinge of Europe, 1081–1797* (1974), *Plagues and Peoples* (1976), *The Pursuit of Power: Technology, Armed Force, and Society since AD 1000* (1982), *Arnold J. Toynbee: A Life* (1989), *Keeping Together in Time: Dance and Drill in Human History* (1995).

Stephen Mennell (1944–) was a sociologist at the University of Exeter, England (subsequently Professor at Monash University, Australia and University College Dublin, Ireland). He is General Editor of *Collected Works of Norbert Elias*, 18 vols (2006–2014). His published works include *All Manners of Food: Eating and Taste in England and France from the Middle Ages to the Present* (1985) and *The American Civilizing Process* (2007).

Immanuel Wallerstein (1930–2019) was Professor of Sociology, and Director of the Fernand Braudel Center for the Study of Economies, Historical Systems and Civilisation, State University of New York, Binghamton, NY. Later President of the International Sociological Association. He is most famous for the 'world-systems theory' expounded in the four volumes of *The Modern World-System* (1974–2011).

Other Participants in the Conference

Sergio Bertelli (1928–2015), Professor of History, University of Florence.

Maarten C. Brands (1933–2018), Professor of Modern History, University of Amsterdam.

Christien Brinkgreve (1949–), Sociologist, University of Amsterdam, later Professor of Women's Studies, University of Nijmegen and of Sociology, Utrecht University.

André Burguière (1938–) Historian of *mentalités*, Directeur des Études, École des hautes études en sciences sociales, Paris.

Gabriele Calvi (1925–2015), Psychologist, then Professor of Political Science, University of Florence, and later Pavia.

Alessandro Cavalli (1939–), Professor of Sociology, University of Pavia.

Eric Dunning (1936–2019), Sociologist (notably of sport), later Professor, University of Leicester.

Georg Elwert (1947–2005), Sociologist and anthropologist; in 1984 *Privatdozent* at the Universität Bielefeld, later Professor of Ethnology and Social Anthropology at the Free University Berlin.

Peter R. Gleichmann (1932–2006) Architect and Professor of Sociology, University of Hannover.

Maria Goudsblom (1936–2009), Social psychologist, Amsterdam.

Franz-Xaver Kaufmann (1932–) Professor of Social Policy and Sociology, Universität Bielefeld, and one of the Directors of its Center for Interdisciplinary Research (ZiF), 1979–1983.

Richard Kilminster (1942–), Sociologist, University of Leeds.

Helmut Georg Koenigsberger (1918–2014) Professor of History, King's College, University of London.

Hermann Korte (1937–), Professor of Sociology, Ruhr-Universät Bochum, subsequently Professor, Hamburg University. Member of the Board of the Norbert Elias Foundation, Amsterdam.

Slawomir Magala (1950–) Philologist, philosopher of science and sociologist of culture and organisation, then at the University of Poznan, later Professor of Cross-cultural Management at the Rotterdam School of Management, Erasmus University Rotterdam.

Georg Stauth (1942–) Islamist and developmental sociologist, Bielefeld and many other universities internationally.

Abraham van Stolk (1941–1997) Sociologist, University of Amsterdam; Member of the Board of the Norbert Elias Foundation, Amsterdam.

Abram de Swaan (1942–), Professor of Sociology, later University Professor, University of Amsterdam.

Nico Wilterdink (1946–), Sociologist, later Professor of Cultural Sociology, University of Amsterdam.

Cas Wouters (1943–) Sociologist, University of Utrecht.

Introduction

Artur Bogner and Stephen Mennell

The long-term development of humanity or human societies was a central interest for the emerging discipline of sociology. It was a major preoccupation for Auguste Comte, Herbert Spencer, Karl Marx, Emile Durkheim and Max Weber, and of lesser figures in the nineteenth and early twentieth centuries such as L. T. Hobhouse in Britain, W. G. Sumner in the USA and Ferdinand Tönnies, Ernst Troeltsch and Franz Oppenheimer among others in Germany. After World War II, however, sociology underwent what Norbert Elias called a 'retreat of sociologists into the present'.[1] Most sociologists became more concerned with studying the local or national societies in which they themselves lived, especially 'social problems' that

[1] Norbert Elias, 'The retreat of sociologists into the present', *Essays III: On Sociology and the Humanities* (Dublin: UCD Press 2009 [Collected Works, vol. 16]), pp. 107–26; the first version of this essay was published in German in 1983.

A. Bogner
Berlin, Germany
e-mail: Artur.Bogner@web.de

S. Mennell (✉)
School of Sociology, University College Dublin, Dublin, Ireland
e-mail: Stephen.Mennell@ucd.ie

A. Bogner, S. Mennell (eds.), *Civilisations, Civilising Processes and Modernity – A Debate*, Palgrave Studies on Norbert Elias,
https://doi.org/10.1007/978-3-030-80379-7_1

they hoped their research could serve to ameliorate. Elias's dissatisfaction with this trend stemmed not from any lack of interest in present-day society and social welfare, still less from a simple interest in history, but from his belief that *time* was an essential axis in any adequate sociological explanation whether of the present or the past. Knowing the past was also essential to any insight into what the future might hold. And he wanted to get away from the use of static concepts and the search for static structures and to foster a more thoroughly processual way of thinking.[2]

On 14–17 June 1984, Elias convened a conference in Bielefeld to which he invited an international group of about 30 scholars who shared his wish to revive the older interest in the investigation of very long-term processes of social development. Among them were some internationally celebrated names: apart from Elias himself, Immanuel Wallerstein, William H. McNeill, Keith Hopkins, Johann P. Arnason and Johan Goudsblom. Michel Foucault and Darcy Ribeiro had accepted the invitation but were in the event unable to attend the symposium.[3]

Many of the participants were meeting each other for the first time, and this was the only occasion when all of them met together to debate their work among themselves. Therein lies the significance and interest of this book. In it, we belatedly make available in one place not only seven of the main papers presented by participants, but also for the first time the transcript of the discussion between them over the three days.[4] It is fascinating

[2] Elias developed a whole sociological theory of knowledge and the sciences although, apart from passing references to 'involvement' and 'detachment', this did not arise prominently at the Bielefeld conference. See *Involvement and Detachment* (Dublin: UCD Press, 2007 [Collected Works, vol. 8]), and *Essays I: On the Sociology of Knowledge and the Sciences* (Dublin: UCD Press, 2009 [Collected Works, vol. 14]).

[3] The list of those who had been invited but did not participate included Pierre Bourdieu, Wolfram Eberhard, Michel Foucault, Clifford Geertz, Jürgen Kocka, Barrington Moore, Darcy Ribeiro, Marshall D. Sahlins, Theda Skocpol and Fei Xiao-tung.

[4] Tape recordings rarely capture every word of such a discussion. Not only are there gaps, but of course speakers habitually make many 'false starts' and other errors that they would not make in writing. We have therefore gently edited and 'tidied up' the transcript, originally made by Ms J. Gorney while always remaining true to the spirit of what was said. Nor was everyone speaking their native language. We have broadly followed the principles set out in the 'Note on editorial policy' in the final volume of the Collected Works of Norbert Elias, *Supplements and Index* (Dublin: UCD Press, 2014), pp. ix–xiv. The original transcript can be examined in the Deutsches Literaturarchiv, Marbach am Neckar. As editors, we have provided all the notes to the transcript.

now to look at their distinct perspectives. Although the participants were invited because of their broader interest in long-term processes, the title Elias chose for the conference was 'Civilisations and Theories of Civilising Processes: Comparative Perspectives', somewhat slanted towards his own central theory of civilising processes, and thus ensuring that the discussion repeatedly returned to his own territory.

The Flight from the Past?

Was Elias right in his complaint about the retreat of sociologists into the present? Certainly, there seemed to be something in the mood of the times, the *Zeitgeist*, the spirit of the age in the post-war decades. In many countries, there was a general focus on 'fixing things' in a practical way, repairing the ravages of war and improving social conditions. There was revulsion at the recent wars and brutalities both of fascist and communist regimes, and an associated distrust of sweeping historical models; even Arnold Toynbee's monumental twelve-volume *A Study of History*, completed in 1961, after causing a little stir, sank almost without trace.[5] There was widespread fear of nuclear war between the USSR and USA. Many commentators of the time claimed to detect a general sense of pessimism, and the mood affected more than sociology.

For the former great powers of Western Europe, after World War I came the apparent end of their global hegemony, especially through a changed power balance with the USA and also in relation to Latin America (Argentina and Brazil for example); and very shortly after World War II came the decolonisation of their former colonies in Africa and Asia. Any new theory of long-term trends in history would not be able to present a vision of the future that would be as magnificent and as flattering as their immediate past, when they had been global empires or what came to be called superpowers. In this role, they were replaced by the USA and the Soviet Union. Such a flattering vision of their future could, so it seemed, only assume the more abstract form of either capitalism or socialism as the final stage of history.

[5] Arnold J. Toynbee, *A Study of History*, 12 vols (Oxford: Oxford University Press, 1934–1961).

Besides these elements of the *Zeitgeist* in the background, there were some specific intellectual influences that helped to promote and justify the trend towards present-centredness among sociologists. For the philosopher Karl Popper (1902–1994), notably influential especially in Britain, it was less a flight from the past than a flight from the future—any prediction being suspected of tending towards dangerous 'historical prophecy', identifying the pursuit of 'Inexorable Laws of Human Destiny' with totalitarianism, whether fascist or communist.[6] He favoured 'piecemeal social engineering', which chimed with the more pragmatic, today-centred interests of the large majority of sociologists. Other rather different philosophies with much the same effect included phenomenology—Alfred Schütz (1899–1959) had considerable posthumous influence in the 1960s and 1970s—and the Wittgensteinians. Notable among the latter was Peter Winch (1958), who drew upon Wittgenstein and analytic philosophy in his attempt to discredit the whole possibility of there being such a thing as a social science.[7] Between them, phenomenology and Wittgenstein spawned the ahistorical ultra-micro-sociology known as ethnomethodology. And among the broader public at least, the 'social psychology' of smaller groups had—perhaps for decades—been the most popular and most publicised speciality inside the discipline, in line with a fast-growing public interest in subjects such as the relations between management and employees in the workplace, in 'prejudice' or stereotypes in the relations between different social, 'racial' or ethnic groupings. Besides, there were no doubt much more mundane forces driving the retreat into the present, notably that funding for research—from governments or foundations—was more easily granted for practical research of modest scope. This was the age of the welfare state.

Sociology first took shape against a very different background. As a specialist discipline, it was not established until significantly later than the other human sciences, especially economics and history; the first chairs and departments of sociology in universities date only from the end of the nineteenth century and the early twentieth century. Moreover, for many of its early exponents, the boundaries between sociology, history, economics and anthropology were far more permeable than they would later become. It was thus not inappropriate that although a majority of the

[6] Karl R. Popper, *The Poverty of Historicism* (London: Routledge & Kegan Paul, 1957), p. iii; see also *The Open Society and Its Enemies* (London, Routledge & Kegan Paul, 1945).

[7] Peter Winch, *The Idea of a Social Science* (London: Routledge & Kegan Paul, 1958).

participants in the 1984 Bielefeld conference were sociologists, their interests were notably interdisciplinary.

The emergence of the social sciences as a whole may be seen as a response—at first a very slow response, but later accelerating—to an increasing awareness in Europe of the global scale and pervasiveness of changes in the structures of societies. A spreading awareness of social change can perhaps be traced back to the Reformation, the voyages of discovery and opening up of new sea routes to other continents, and the printing press. The 'Gutenberg Revolution' was associated both with translations of the Bible into the vernacular languages first of Europe (and, after 1800, of many other peoples and countries) and with the stirrings of the modern natural sciences.[8] From the Enlightenment the pace quickened, the political movements before and after the American and French Revolutions bringing transformations in the state apparatus. The major changes also included increasing schooling and literacy for broader sections of the home population, while colonial administrations and missionary movements brought about revolutionary transformations in the economies, educational systems, and later also the health systems of colonial areas outside Europe. With some cultural lag, all this was also associated with the gradual breaking of the church's monopoly over the means of orientation.[9] Last but not least was the accelerating pace of industrialisation in Europe in the nineteenth century. And not just there: European overseas expansion changed not only the economic systems and relations, but also stalled or refashioned the 'endogenous', 'native' processes of state formation and political development in the overseas colonies, and the 'Global South' more generally.

In this perspective, it is easy to see why seeking to understand processes of historical development, along with theories of 'progress' that were to be

[8] See amongst others Niall Ferguson, *Civilization: The Six Killer Apps of Western Power* (London: Penguin, 2012), pp. 60–9.

[9] This was not at all sudden: for example the published reports of missionaries about their adventures and feats overseas as well as biographies of them or the local converts were among the most widely read products of the printing press, for example in the German-speaking countries in the nineteenth century. See Artur Bogner, Bernd Holtwick and Hartmann Tyrell (eds), *Weltmission und religiöse Organisationen* (Würzburg: Ergon, 2004). On the momentous and fateful *intertwining and synergy* of overseas discoveries, *socio-institutional changes*, modern science, political expansion, and economic growth see the forceful argumentation and concise summary by Yuval Noah Harari, *Sapiens: A Brief History of Humankind* (London: Vintage, 2014 [2011]), Chaps. 15 and 16. On this synergy, see also the partly similar but less coherent or systematic arguments in Ferguson, *Civilization*.

such a problematic legacy later, were guiding principles of early social or socio-cultural scientists. Closely associated was the idea of 'civilisation'. This concept had gradually emerged in the eighteenth century to denote the *process* of people becoming more 'civil', or 'polite'.[10] From being a mark of upper-class manners, by the mid-nineteenth century it had become a badge of the supposed superiority of the upper classes, especially the aristocracy, over lower classes in European countries—and then of proclaimed European pre-eminence over all others, notably their colonised peoples. Europeans also lost any sense of 'civilisation' having a history, of having developed gradually over many generations, and experienced it as a static manifestation of what they had come to see as their inherent superiority. Well aware of how the concept had come to have these connotations, Norbert Elias had nevertheless attempted to revive the sense of civil*isation* as a long-term process, while still using the same word.[11] As will be seen, these historic connotations and role of the word were still a source of controversy in the discussions at the 1984 Bielefeld conference.

In spite of such or similar difficulties, historically orientated sociology had never disappeared in the twentieth century, and indeed the minority of its proponents often enjoyed considerable prestige as 'theorists'. German sociology of the Weimar period—of which Norbert Elias, the host of the Bielefeld conference ranked in 1984 as one of the last survivors—was dominated by what may be seen as variants of historical sociology, represented by Max and Alfred Weber, Ferdinand Tönnies, Ernst Troeltsch, Franz Oppenheimer, Karl Mannheim, Emil Lederer, Karl

[10] See Norbert Elias, *On the Process of Civilisation: Sociogenetic and Psychogenetic Investigations* (Dublin: UCD Press, 2012 [Collected Works, vol. 3]), pp. 13–67; Bruce Mazlish, *Civilization and its Contents* (Stanford, CA: Stanford University Press, 2004).

[11] Right at the end of his life, Elias acknowledged that this question had led to endless misunderstanding of his work, but said that in the 1930s he had tried and failed to find an alternative word that would capture the full range of meanings he wanted to explore; see Elias, 'What I mean by civilisation: reply to Hans Peter Duerr', *Essays II: On Civilising Processes, State Formation and National Identity* (Dublin: UCD Press, 2008 [Collected Works, vol. 15]), pp. 8–13. Stephen Mennell has suggested that the familiar concept of deferred gratification, broadly interpreted to encompass social and cultural as well as psychological aspects, might have captured most though not all of the senses that Elias wanted to capture. Another facet or aspect of his concept is what Sigmund Freud had earlier, for the individual, called 'sublimation'; see Mennell, 'Childhood and society: civilisation as deferred gratification', *Sozialwissenschaftliche Literatur Rundschau (SLR)*, Heft 77, 2018, pp. 92–9.

A. Wittfogel, Franz Borkenau and others.[12] Many notable American sociologists, social theorists or social philosophers of the post-war years had their roots in pre-war German sociology. Parts of the work of Reinhard Bendix and Edward Shils, for example, retained an historical slant.[13] And Talcott Parsons, who dominated sociological theory from the 1940s to the 1960s, had gained his doctorate in Heidelberg in the 1920s. After leading a retreat into the static models of his 'structural-functionalism', Parsons's concern with 'modernisation' led him into something of a revival in evolutionary theorising in the later 1960s and 1970s. This overlapped with a growing wider interest in Max Weber's sociological writings on economic history and the history of world religions. In 1963, a famous seminar on social evolution took place at Harvard with Parsons, Robert Bellah and Shmuel Eisenstadt.[14] Parsons published two small books on the long-term evolution and modernisation of societies.[15] In themselves, these books amounted theoretically to little more than a revival of Herbert Spencer's model of social evolution from Victorian times, complete with the USA as the 'lead society'. Another major contribution in English to the revival of social evolutionism, with a distinctly Spencerian red in tooth and claw flavour, was the work of W. G. Runciman, of which the first volume had been published just before the Bielefeld conference.[16] The work

[12] For an excellent survey, see Volker Kruse, 'Von der historischen Nationalökonomie zur historischen Soziologie: Ein Paradigmenwechsel in den deutschen Sozialwissenschaften um 1900', *Zeitschrift für Soziologie* 19: 3 (1990), pp. 149–65.

[13] Reinhard Bendix, *Work and Authority in Industry: Managerial Ideologies in the Course of Industrialization* (New York, Wiley, 1956), and *Nation-Building & Citizenship: Studies of Our Changing Social Order* (Berkeley, CA: University of California Press, 1964); *Kings or People: Power and the Mandate to Rule* (Berkeley, CA: University of California Press, 1978). Edward Shils, *The Constitution of Society* (Chicago: University of Chicago Press, 1972).

[14] See Talcott Parsons, 'Evolutionary universals in society'; Robert N. Bellah, 'Religious evolution'; S. N. Eisenstadt, 'Social change, differentiation and evolution', *American Sociological Review* 29: 3 (1964), pp. 375–86, 358–74 and pp. 339–57 respectively.

[15] Talcott Parsons, *Societies: Evolutionary and Comparative Perspectives* and *The System of Modern Societies* (Englewood Cliffs, NJ: Prentice-Hall, 1966 and 1971). It should be noted that Elias explicitly preferred to reserve the term 'evolution' for biological processes, always using 'development' for social processes, because the differences between the two are very marked—notably that social processes are generally reversible, while biological ones are not. This distinction is not, however, consistently observed among sociologists; and in the United States, some writers have used the terms the other way round.

[16] W. G. Runciman, *A Treatise on Social Theory*, vol. 1, *The Methodology of Social Theory*; vol. 2, *Substantive Social Theory*; vol. 3, *Applied Social Theory* (Cambridge: Cambridge

of Eisenstadt, however, has always been recognised as a milestone in the renewal of a large-scale empirical historical sociology.[17]

In Germany, the revival of functionalist evolutionism was represented both by Niklas Luhmann (1927–1998) and Jürgen Habermas (1929–), the two major social theorists who emerged after the 'cultural revolution' and the student revolts of the 1960s; though they were often at odds in other respects, they were both strongly influenced by the theories of socio-cultural evolution and modernisation associated with Parsons, Bellah and Eisenstadt. Habermas—the leader of the second generation of the 'Frankfurt School' in sociology—and some of his renowned students or sympathisers (such as Klaus Eder, Rainer Döbert, Claus Offe and Johann P. Arnason[18]) tried to combine some form of systems theory or structural functionalism with neo-Marxist research questions. The legacy of Marxism and German historical sociology (including Max Weber's theories of collective or institutional 'rationalisation') also ensured that the passage of time, and more precisely the dimension of temporal structuredness and sequential patterns of long-term change, did not disappear from sociological thinking. The inheritance and resurgence of Marxism, stronger and more deeply rooted in Germany than in the USA, also meant that there was a considerable reception in the 1970s and 1980s for the writings on the history or historical sociology of state formation by 'left' scholars such as Immanuel Wallerstein, Perry Anderson, Barrington Moore Jr. and Moore's influential student Charles Tilly. And Marxian theories of *dependencia*—that is of *dependent* economic and socio-economic development in the (formerly) colonised countries or poor regions of the world were comparatively far more respected in what was then called the 'Third World', including Latin America.[19] In most other disciplines and in other

University Press, 1983, 1989, 1997 respectively); *The Theory of Cultural and Social Selection* (Cambridge: Cambridge University Press, 2009).

[17] For a notable early work, see S. N. Eisenstadt, *The Political Systems of Empires* (New York: Free Press, 1963).

[18] Johann Arnason, participant of the ZiF conference and author of a chapter in this book, is a former doctorand of Habermas.

[19] Immanuel Wallerstein, *The Modern World-System*, vol. 1, *Capitalist Agriculture and the Origins of the European World-Economy in the Sixteenth Century* (New York: Academic Press, 1974) and subsequent volumes; Perry Anderson, *Passages from Antiquity to Feudalism* and *Lineages of the Absolutist State* (London: New Left Books, 1974); Barrington Moore Jr., *Social Origins of Dictatorship and Democracy* (Boston: Beacon Press, 1966); Charles Tilly (ed.), *The Formation of National States in Western Europe* (Princeton, NJ: Princeton University Press, 1974). During the 1970s, Dieter Senghaas, a pioneer and for about two

Western countries, neo-Marxism or similar theories of 'socio-cultural evolution' were not too popular, at least not in the so-called 'Western world'. This was, after all, also the age of the Cold War.

So, in sociology, the picture is mixed. Nevertheless, Elias's claim that there had been a 'retreat into the present' remains broadly fair when compared with academic sociology in the pre-war era. Numerically speaking, a large majority of post-war sociologists pursued present-centred, ahistorical research, often funded by governments and foundations. If sociologists in this period were interested in social development, the word usually applied to studies in the 'developing countries', carrying with it the subliminal message that 'development' in the industrial societies was done and finished, that it remained only to be exported to the 'Third World'.

Of course, this was not confined to sociology but rather a part of a much broader trend in the social sciences with deep roots. Neoclassical economics came to pay little attention to long-term processes, focusing increasingly on timeless, nomothetic theories about the 'laws' of supply and demand in present-day markets. Marx had spotted the tendency much earlier:

> When the economists say that present-day relations—the relations of bourgeois production—are natural, they imply that these are the relations in which wealth is created and productive forces developed in conformity with the laws of nature. ... They are the eternal laws which must always govern society. Thus there has been history, but there no longer is any.[20]

decades the doyen of development research in political science and sociology in Germany, did much to promote the reception of theoretical works from the Global South. His publications are typical of the friendly reception, in Germany, of such analyses of long-term changes of societal structures—see Dieter Senghaas (ed.), *Kapitalistische Weltökonomie: Kontroversen über ihren Ursprung und ihre Entwicklungsdynamik* (Frankfurt: Suhrkamp, 1979), *Imperialismus und strukturelle Gewalt* (Frankfurt: Suhrkamp, 1972) and *Peripherer Kapitalismus: Analysen über Abhängigkeit und Unterentwicklung* (Frankfurt: Suhrkamp, 1974).

[20] Karl Marx, *The Poverty of Philosophy* (Moscow: Foreign Languages Publishing House, n.d.), p. 135. This observation may perhaps be seen as anticipating Francis Fukuyama's explicitly Hegelian conception of an 'end of history', after the collapse of communism: see *The End of History and the Last Man* (London: Hamish Hamilton, 1992). His hotly debated essay, 'The End of History' first appeared, in the journal *The National Interest*, as early as the summer of 1989.

Writers such as Friedrich Hayek (1944) came to have a similar impact in economics to that of Popper on sociology, in identifying theories of long-term development—and any planning to bring it about—with a drift to authoritarianism. Economic history gradually came to occupy a smaller part of the economics syllabus from the 1960s onwards.

Even historians were not immune. Historians, needless to say, could hardly abandon history. But there was a shortening of time spans. Perhaps it was increasing professionalisation of the discipline that propelled a retreat to the respectability of studies of short periods—or at most, text-books on one century or so, with a rare daring excursus to such stretches as the long sixteenth century (1500–1640). Here, too, there were important exceptions. The French *Annales* school—established by Lucien Febvre and Marc Bloch in the 1920s and carried forward to the present day by Fernand Braudel, Emmanuel LeRoy Ladurie, Roger Chartier among many distinguished members—became widely influential. In the USA, one historian who took part in the Bielefeld conference, William McNeill, had defied the general disapproval in his profession of the large-scale and long-term to write books as ambitious as *The Rise of the West*, and perhaps despite some frowns (it is said) he was elected President of the American Historical Association for 1985.[21] But even he, as he makes clear in the discussion, continued to share in the historians' general suspicion of theory, especially theories of long-term historical processes. One of the few prominent historians to have a positive taste for theory is Peter Burke, who has repeatedly shown the relevance of social theory to history and history to social theory.[22]

[21] William H. McNeill, *The Rise of the West: A History of the Human Community* (Chicago, IL: University of Chicago Press, 1963). Equally ambitious books of global history include: *A World History* (New York: Oxford University Press, 1967), *Plagues and Peoples* (Garden City, NY: Anchor, 1976), *The Pursuit of Power: Technology, Armed Force and Society since AD 1000* (Chicago: University of Chicago Press, 1982) and, with John R. McNeill, *The Human Web: A Bird's Eye View of World History* (New York, W. W. Norton, 2003).

[22] Burke, Peter, *Sociology and History* (London: Allen & Unwin, 1980); rev. edn. *History and Social Theory* (Cambridge: Polity, 1992); 3rd edn, same publisher 2005.

The Bielefeld Conference

When he convened the 1984 conference, Norbert Elias was in his final year as a Resident Fellow (an informal title) at the Zentrum für interdisziplinäre Forschung (ZiF) in Bielefeld University, a position he had held since 1978.[23] The university had been founded only in 1969, and was organised in distinctive way; it emphasised interdisciplinary research and the close connection between research and teaching, with most academic staff teaching in their own areas of research. By the late 1970s, it already had large and internationally renowned departments both of sociology and of history, making it especially congenial for Elias, and today it is home to the Bielefeld Graduate School in History and Sociology.

Niklas Luhmann (1927–1998), who was to become the most reputed sociological theorist in Germany during the 1980s–1990s, joined the Fakultät für Soziologie at Bielefeld University from the beginning (he was the first full university professor of this faculty), but he seems not to have been one of those with whom Elias had much contact. His highly abstract systems theory and his attempt to eliminate the individual human being from his theorising put him poles apart from Elias. The history department was more congenial. Under the leadership especially of Hans-Ulrich Wehler (1931–2014) and Jürgen Kocka (1941–), it developed into the home of a new school of structural, social or 'societal' history ('*Gesellschaftsgeschichte*'), which became known as 'the Bielefeld School of History'. Wehler had helped to introduce or reintroduce Elias to contemporary German historians[24] after Elias's principal work, *Über den Prozess der Zivilisation*, was reissued in 1969, though Wehler too was not present at the 1984 conference.

Another conspicuous absentee was Michel Foucault, who had been expected. In accepting Elias's invitation, he had written

[23] He was the only such Fellow to be resident for such a long period. The ZiF, situated right in the Teutoburg Forest, and close to, but not actually on the main campus, is one of Europe's distinguished 'institutes of advanced study', at which Resident Fellows are usually appointed for only one year. It suited him well, when already in his 80s, to have a flat next to the ZiF's indoor swimming pool and adjacent to the forest. His private home was by then in Amsterdam. In the years 1982–1984 Elias spent about half his time in his apartment at Bielefeld and half at his home in Amsterdam.

[24] See Elias's contributions in two edited volumes edited by Wehler, *Geschichte und Soziologie* (Köln: Kiepenheuer & Witsch, 1972); *Soziologie und Psychoanalyse* (Stuttgart: Kohlhammer, 1972); Wehler, already prominent then, soon became one of the most celebrated German historians.

> Thank you very much for your letter. You know how much I admire your work and all that I owe to your research. This is to say how happy I shall be to be able to meet you in the framework you propose.[25]

He went on to say that he was currently working on 'techniques of the self' in Antiquity.[26]

So it is sad to read what Elias said in welcoming the participants to the conference:

> I am afraid Professor Foucault has not yet arrived. I still hope that he will arrive. He has not said that he will not come. He has said he would come. What links his interest and mine, I think, is the preoccupation with the relationship between social constraints and self-constraints. He approaches it in a somewhat different manner from myself but I can say for myself that the whole problem of constraints in theory and practice is an interesting point, perhaps a little less regarded because we don't like to know of constraints, we like to believe that we are all free to do what we are like, what we like. But we all know that this is not the case. So research into constraints, social as well as self-constraints is a fruitful field, in a long-term perspective. And I still hope that Professor Foucault will come and give us the benefit of his knowledge.

Only a few days after the conference did we learn the reason for Foucault's absence: he died on 25 June 1984.

Also missing was Darcy Ribeiro (1922–1997), the Brazilian anthropologist, sociologist, and politician, whose 1968 book *The Civilizational Process*[27]—especially under its title in English—sounds very close to Elias's own (otherwise known as *The Civilizing Process*). But in fact, the resemblance was not as great as the titles suggested. The main influence on Ribeiro's thinking about the long-term development of human society was the neo-evolutionist theory of the American anthropologists Leslie White and Julian Steward, and the British Marxist archaeologist V. Gordon

[25] Letter, in French, dated 9 July 1982 from the Collège de France (our translation—eds).

[26] Foucault would have been referring to his *History of Sexuality*, and perhaps to volumes 2 and 3 especially: *The Use of Pleasure* (London: Penguin, 1984) and *The Care of the Self* (London: Allen Lane, 1988). Both were first published in French in 1984.

[27] Darcy Ribeiro, *O processo civilizatório: etapas da evolução sócio-cultural* (Rio de Janeiro: Civilizacão Brasileira, 1968); English translation by Betty J. Meggers: *The Civilizational Process* (Washington, DC: Smithsonian Institution Press, 1968).

Childe. The stages of the 'civilisatory process', equated here with the sequential order of societal and with economic formations in the history of humankind, were marked by a series of technological breakthroughs, or revolutions.

A full list of participants in the conference can be found on pp. ix–xii above. Apart from the papers by Evers, Wallerstein, Elias, McNeill, Hopkins, Arnason and Goudsblom, a few papers were circulated by other participants—Georg Elwert, Stephen Mennell and Georg Stauth, for example[28]—but not presented orally. We have included here the seven papers that were delivered orally and largely shaped the discussions. The proceedings were recorded and transcribed, and it was always intended that they be published in a book. That did not happen at the time—Elias became preoccupied with a rush of other books and articles that appeared in the remaining six years of his life and immediately after his death.[29] It is in many respects the discussion among these scholars that is most revealing. Readers should not expect much consensus or clarity to emerge. It will be found that the main contributors sometimes talked past each other. Most of them were meeting each other for the first time, and there was a lot to digest. For instance, William H. McNeill admitted (p. XXX) that his training as an historian gave him a different cast of mind from that of the sociologists and, moreover, that he encountered Elias's ideas only after being invited to the conference. He expressly admired *The Court Society* (as historians quite often do) but was sceptical about *On the Process of Civilisation*. Interestingly, though, for the rest of his life after the conference McNeill maintained a long correspondence with Johan Goudsblom—who was one of the most influential advocates of Elias's theories in European sociology and we know from him that McNeill's scepticism never abated.

[28] Revised versions of Elwert's and Mennell's papers were published afterwards: see Georg Elwert, 'Ausdehnung der Käuflichkeit und Einbettung der Wirtschaft: Markt und Moralökonomie', in Klaus Heinemann (ed.), *Soziologie wirtschaftlichen Handelns* (*Kölner Zeitschrift für Soziologie und Sozialpsychologie*, Sonderheft 28, 1987), pp. 300–21; and Stephen Mennell, 'On the civilising of appetite', *Theory, Culture and Society* 4: 2–3 (1987, pp. 373–403.

[29] To list only the books: *Quest for Excitement* (with Eric Dunning, in English, 1986), *The Society of Individuals* (in German 1987), *Studies on the Germans* (in German, 1989), *Mozart: The Sociology of a Genius* (in German 1991), and *The Symbol Theory* (in English, 1991). These can be found in volumes 10, 11, 12 and 13 respectively of his Collected Works in English.

With the generous and energetic support of the directorate of the ZiF, particularly Professor Franz-Xaver Kaufmann, who was then one of its members, and of Dr Gerhard Sprenger, the manager of the ZiF, Elias obtained funding for the conference from the Volkswagen Foundation. A major role in organising the conference was played by Artur Bogner, who was Elias's private secretary during the period 1982–1984 and who had been officially entrusted by the ZiF with the preparation and organising of this conference. He wishes to express his deep gratitude for the generous support and advice of Professor Kaufmann and the late Professor Georg Elwert.

Subsequent Trends

Whether by coincidence or not, the conference proved to be a harbinger of a revival of macro-history and large-scale historical sociology. Two years later, there appeared the first volume of Michael Mann's five-volume 'history of power', which stretches from 'the beginning' to the present day.[30] Still more ambitious was the 'big history' movement, led by David Christian, which sought to set the development of human society in the context of the development of the universe, no less—cosmology, geology, physics, biology and ecology.[31] (This enterprise, worthy of Auguste Comte and Herbert Spencer, was to make the famous 'Plato to NATO' undergraduate survey courses look quite small-scale.) More significant, however, has been the emergence of much wider *popular* interest in studies of history on the large scale. One widely read example is by Jared Diamond: his *Guns, Germs and Steel* is entitled 'a short history of everybody for the last 13,000 years'.[32] Another important bestseller is Yuval Noah Harari's *Sapiens*.[33] In this 'brief history of humankind', he points out that from a bird's eye or satellite's view it is very clear that the main long-term current of history has been and continues in the direction of closer contact and interaction, in the sense of a unification of all human communities—that is their confluence into a single web of communication

[30] Michael Mann, *The Sources of Social Power*, vol. I, *From the Beginning to AD 1760* (Cambridge: Cambridge University Press, 1986).

[31] David Christian, *Maps of Time: An Introduction to Big History* (Berkeley, CA: University of California Press, 2004).

[32] Diamond, Jared, *Guns, Germs and Steel: A Short History of Everybody for the Last 13,000 Years* (London: Chatto & Windus, 1997). See also *Collapse: How Societies Choose to Fail or Survive* (London: Viking, 2005).

[33] Harari, *Sapiens*.

and interdependencies. Meanwhile this does not imply, certainly not so far, a clearly increasing accord and homogeneity in the political and cultural dimensions. This discrepancy was also one of the threads running through the discussions at the 1984 conference. And it is a useful note on which to end this Introduction: by analogy with meteorology, it may be easier to see and predict the broader and longer-term trends in history than to make detailed short-term predictions. We are able to predict with a relatively high degree of confidence that downpours and thunderstorms will occur in certain regions and seasons, but we cannot predict specifically where and when there will be floods or flashes of lightning.

The Civilising Process: World Figuration or World System?

Hans-Dieter Evers

I

Theories of civilisation have tended to stress the *longue durée* of historical developments.[1] The civilising process, as defined by Alfred Weber, is a directed, irreversible process of increasing consciousness of the surrounding world (*Bewusstseinsentwicklung und Durchreflektion des Daseins*).[2] This in turn leads to a systematisation of knowledge and its practical application.[3] This process can therefore be transferred from one 'civilisation' (*Geschichtskörper*) to another, without, however, necessarily transcending the whole world. A certain social formation (*Gesellschaftsprozess*) might speed up or delay the process of civilisation,

[1] Fernand Braudel, *On History* (London: Weidenfeld & Nicolson, 1980).
[2] Alfred Weber, *Ideen zur Staats- und Kultursoziologie* (Karlsruhe: Braun, 1927).
[3] Alfred Weber, *Prinzipien der Geschichts- und Kultursoziologie* (Munich: Piper, 1951), p. 72.

H.-D. Evers (✉)
Berlin, Germany

A. Bogner, S. Mennell (eds.), *Civilisations, Civilising Processes and Modernity – A Debate*, Palgrave Studies on Norbert Elias,
https://doi.org/10.1007/978-3-030-80379-7_2

and the cultural process (*Kulturbewegung*) of unpredictable creativity might throw it completely off course. The analysis of the changing 'constellation' of the three processes of social formation, culture and civilisation provides, according to Alfred Weber, a key for an understanding of world history.[4] The civilising process remains, however, the driving evolutionary force, but the result is not necessarily a unified 'world civilisation' or 'world system', nor is it implied that all societies have to go through a similar cycle of rise and decline.

Norbert Elias's theory of civilisation steers clear of evolutionary determinism.[5] His central concept of figuration is fundamentally different from the concept of constellation or social system. The European court of the sixteenth and seventeenth centuries is such a figuration. Quite appropriately Elias uses court dancing to illustrate his concept: a multitude of actors change positions and partners, but the figuration (type of dance) remains visibly the same.[6]

Here the crucial question arises how the figuration takes place—or, to stay with his example, how does one dance style develop and change into another one? The answer is, through a civilising process of increasing control of feelings (*Affektkontrolle*), sophistication of manners (*Verfeinerung der Sitten*), control of violence (among other things) and the institutionalisation of these refined manners, of inner control, in the form of the state. Eventually, the absolutist and modern state emerges.[7] Elias does not pursue his analysis much beyond this stage of state formation, but towards the end of his study, he detects the trend towards a new figuration (*Verflechtung*), 'that spans the whole inhabited world'.[8]

In Wallerstein's or Luhmann's systems theory, the current world is already one social system.[9] An earlier plurality of small societies and world systems in

[4] A. Weber, *Ideen*.

[5] Norbert Elias, *On the Process of Civilisation: Sociogenetic and Psychogenetic Investigations* (Dublin: UCD Press, 2012 [Collected Works, vol. 3]).

[6] Norbert Elias, *The Society of Individuals* (Dublin: UCD Press, 2010 [Collected Works, vol. 10]), p. 23. (This book was not published, in German, until 1987, but this first section of it had been written as early as 1939, and in conversation Elias often used the image of dance.)—eds.

[7] Norbert Elias, *The Court Society* (Dublin: UCD Press, 2006 [Collected Works, vol. 2]).

[8] Elias, *On the Process of Civilisation*, p. 480.

[9] Immanuel Wallerstein, *The Modern World-System, vol. 1, Capitalist Agriculture and the Origins of the European World-Economy in the Sixteenth Century* (New York: Academic Press, 1974); Niklas Luhmann, 'The world society as a social system', *International Journal of General Systems*, 8: 2 (1982), pp. 131–8.

the form of empires or large markets has meanwhile been absorbed into a single world system, which is, however, differentiated into a core, a semi-periphery and a periphery. This typical differentiation is, so to speak, the system property of the modern world system. What has led to both the all-encompassing expansion and the differentiation of the modern world system? Wallerstein's analysis is complex, and so far incomplete, as the last few volumes of his work on the modern world system have still to appear. But the major driving force appears to have been the chances for appropriation created by worldwide trade and the logic of capitalist accumulation.

But it may equally well be argued that the process of constituting the modern capitalist world system is nothing but Elias's 'process of civilisation'. In fact, one could also lay claim to the respective arguments made by Max Weber to support this thesis.[10] Weber's inner-worldly asceticism leads to disciplined behaviour in the same way as Elias's inner control of feelings leads to good manners. The outcome of this civilising process is the absolutist state, a rationally organised bureaucracy and a capitalist economy.

Both approaches are distinct, but to my mind compatible. Elias argues much closer to reality. His figurations are not meant to be ideal types, but descriptions of real life, of actually existing *Lebenswelten*. They are, however, not description of historical events but of frameworks in which events take place.

A number of questions remain, however, unanswered if we move to a level of analysis beyond the nation state. The emergence of court society as a figuration is beautifully explained by Elias, but how was it connected with other figurations? What, for example, was the impact of the equally breathtaking civilising process of Islam and the figuration of the Islamic court on parallel developments in Europe?

Can world society be seen as a loose assembly of figurations that arise under certain conditions and decline under others? Or has the civilising process that once started in Central Europe spread across the globe to create a sort of super- or world-figuration called world society, world economy or modern world system? Is the civilising process thus the core of social evolution, as theories of civilisation like to suggest?[11]

[10] Weber, Max *The Protestant Ethic and the Spirit of Capitalism* (London: Allen & Unwin, 1930 [1905–6]).

[11] A. Weber, *Prinzipien*; Darcy Ribeiro, *The Civilizational Process* (Washington DC: Smithsonian Institution Press, 1983 [1968]).

II

The civilising process described by Elias created the European court society, a dominant figuration or in other words a 'civilised' ruling class. But was this civilising process in the sense of inner psychic control and the regulation of force (creation of *gewaltfreier Räume*) ever completed? Certainly not. Elias himself points out that the feudal courts of the twelfth century and even more the absolutist courts later on provided a chance for an increasing emancipation of women, but that *Männerherrschaft* (men's rule) was broken only occasionally.[12] The use of force against women and children in the domestic sphere may in fact have become more severe, the more manners and conduct were regulated in public life.

Also, peasants and serfs did not participate, they were not 'civilised'. Despite peasant wars, it was not the peasantry but the urban trading-cum-handicraft class (the bourgeoisie) that staged the revolutions (French and others) to topple the ruling 'civilised' aristocracy. The European cities, typified by Max Weber as a sort of counter-culture of 'non-legitimate rule', were initially kept apart from feudal civilisation; but eventually they became the heirs of the civilising process of medieval Europe.

This process of *Ausgrenzung* (exclusion) of spheres in which the constraining rules and regulations of the civilising process were not valid seems, indeed, to be part and parcel of the civilising process itself. Perhaps this point can best be illustrated by referring to the world beyond Europe. No sooner had he left his home shores, the chivalrous knight transformed himself into a bloodthirsty conquistador. Gentlemanly behaviour and fair play was not prescribed in the treatment of African slaves or natives in the colonial territories. What was true for the 'civilised world' did not apply to the colonial world. Slave trade, extermination of 'uncivilised' natives and gunboat diplomacy signify areas of unfettered use of brute force, the play of crude emotions. Other civilisations, like the Inca empire in Latin America, or Benin in Africa, were destroyed in a process of decivilisation.

The thesis I wish to pursue here is that any civilising process presupposes a counter-process, that the 'uncivilised world' is not an area 'not-yet-civilised' but one in which 'barbarism' or 'un-civilisation' is created by the civilising process in the core civilisation itself. The civilising process in China created the barbarian as well as the Confucian scholar. Both are part of the same figuration.

[12] Elias, *On the Process of Civilisation*, pp. 278–9.

Dependency theorists have argued for some time that the underdevelopment of the Third World is the result of economic growth and capital accumulation in the central industrialised countries. Others have attempted to prove that in Third World cities a regulated, 'civilised' formal sector of wage labour can only be maintained through the existence of an informal sector with an unlimited supply of cheap labour. From a worldwide perspective it is rather uncertain whether free wage labour for an employer, long regarded as a typical figuration of the European civilising process, is in fact spreading or retracting. The process of civilisation appears to have created its counter-type of unregulated subsistence production for own use, of unwaged housework without regulated working hours and without social security or other blessings of civilisation.[13]

The result of the process of civilisation as analysed by Elias is a modern world system as described by Wallerstein. For the sake of argument, a crude simplification may be permitted: there appears to be a tendency towards the unleashing of uncontrolled force in the periphery, like *violencia* in Latin America or Pol Pot's ravages in Cambodia. This in turn is matched by controlled and regulated force in the Centre, like the deployment of missiles or atomic weapons and the destruction of the natural environment through an efficient road system, regulated rivers and a very civilised industrialised bureaucracy.

III

The civilisation process has continuously produced uncivilisation or barbarism in a dialectical manner. Increasingly, the earlier opposition between culture and nature, between the cultivated world and the jungle has been replaced by the opposition between regulated and unregulated, between system and *Lebenswelt*, between affluence and poverty, between centre and periphery. Evolution is a civilising process, but it appears to be a dialectical process continuously producing its own opposition.

Alfred Weber and Elias point out that social differentiation is accompanying the process of civilisation. In fact, this way new figurations arise. But these figurations should include their own oppositions. The figuration 'court society' is not complete without the peasantry, the figuration of the

[13] Joan Smith, Immanuel Wallerstein and Hans-Dieter Evers (eds), *Households and the World Economy* (Beverly Hills, CA: Sage, 1984).

absolutist state and its highly figurated civilisation includes the destroyed and deformed civilisations in Latin America and Africa.

Evolutionary systems theories insist on social differentiation as the major if not the only evolutionary process. Niklas Luhmann, for instance, argues that European society has evolved into a functionally differentiated system. Because of the novel communicative processes in this extremely complex system, the boundaries can no longer be integrated by common territorial boundaries. A possible sole exception is the modern state—that is, the political subsystem. As the inclusion of all communicative behaviour into one societal system is unavoidable, society becomes a global system. 'A plurality of possible worlds becomes inconceivable'.[14]

In fact, there has been an increase in communication and an expansion of mass culture, foreign trade and multinational or international organisations. In the words of Wilbert E. Moore, 'The world, then, is a singular system ...: to an increasing degree, the life of the individual anywhere is affected by events and processes everywhere'.[15]

If the whole world has indeed evolved into one global social system, evolutionary systems theory would be in a theoretical cul-de-sac. A system, in this case the social system 'world society', would be a system without an environment. The question arises, whether or not a system without an environment can be regarded as a system at all and whether a one-system evolution is possible. In this case, how could the most successful society be selected in an evolutionary process?

Luhmann tries to solve this problem by pointing out that global society is a functionally differentiated system.[16] Evolution consequently becomes more and more dependent on the outcome of evolutionary processes within its internal environment rather than a system environment.

This argument is not entirely convincing. Would it then not be more useful to argue that we do have, indeed, a multitude of social systems or societies, rather than overstressing coherence and interactive communication within a supposedly integrated world social system? Is world society as a social system thinkable at all? If the formation of a world system is a new evolutionary step, can participants think ahead beyond the capabilities of their own social system?

[14] Luhmann, 'The world society', p. 133.

[15] Wilbert E. Moore, 'Global sociology: the world as a singular system', *American Journal of Sociology*, 71: 5 (1966), pp. 475–82.

[16] Luhmann, 'The world society', p. 135.

'Man's ability to participate intelligently in the evolution of his own system is dependent on his ability to perceive the whole', says Wallerstein.[17] But does man [sic] have this ability?

A number of observations suggest that this is not, or not yet, the case. Instead, a new environment is created, perhaps to re-establish the evolutionary potential and the adaptive capacity of an emerging world society. Fascination with space travel and extra-terrestrial beings project an imaginary alternative society into the universe. A frontier for the process of civilisation is opened, in which earthlings are confronted with galactic empires, star wars or images of their own past and future.

Religious revolutionary movements, like the resurgence of fundamentalist Islam in Iran, or the emergence of ethnic identity in a number of world regions from the US to India, could be interpreted as another reaction of the world system, of world society against an evolutionary impasse. By defining a boundary between believers and unbelievers, between one community and another, a multitude of societies are created with reference to others.

IV

So far, we have mainly posed questions rather than proposed solutions. To deal with the whole world as one social system in the strict sense of general systems theory (Luhmann's position) requires the introduction of rather clumsy additional conditions (*Hilfskonstruktionen*). To study the emergence of a modern world system as a long-term historical process of economic integration through increasing world trade (Wallerstein's approach) is closer to reality and consequently more plausible. But in a way, all systems theoreticians seem to get caught in their own creations. The reactions against systemic forces that are becoming very clearly visible from time to time and cannot be overlooked at present disturb the neat images of constructed systems.

A general theory of the process of civilisation (as proposed by Elias) could encompass the changing figurations of an emerging world society and pay heed to the dialectics of civilisation and barbarism. The modern world economy (as studied by Wallerstein) could thus be interpreted as the result of a long-term civilising process of emerging figurations and counter-figurations.

[17] Wallerstein, *Modern World-System*, p. 10.

Discussion of Evers's Paper

Transcript

Elias: I only want to say a very short word. I appreciate very much what Professor Evers has said and with much of it I can agree. But I have to safeguard my own theory against the idea that I am thinking simply in terms of 'civilisation' and 'barbarism'. I think the conceptualisation and too static antagonism and static polarity is a misinterpretation of what I tried to do. It is extremely difficult to think in terms of more or less civilised instead of thinking in terms of 'civilised' and 'barbaric'. But I should like to say briefly without enlarging on it that to the best of my knowledge, there is no group of humans on earth without civilising processes. The most primitive civilising processes are already a human universal in my conception. The nomads, the barbarians, the what you call the counter-figurations have a different level of civilisation. But I would not think that we can use the concept of absolute barbarianism or absolute civilisation.

Wallerstein: Perhaps now I could say a word because I think the issue has been posed. I think though we might distinguish between the two levels of discourse. I think one level of discourse is that of the analyst, a newer use of the term 'civilisation', a new concept of it. And the other is

Transcript (✉)
Berlin, Germany
e-mail: Artur.Bogner@web.de

A. Bogner, S. Mennell (eds.), *Civilisations, Civilising Processes and Modernity – A Debate*, Palgrave Studies on Norbert Elias,
https://doi.org/10.1007/978-3-030-80379-7_3

the description of the concepts that emerge within social structures and the concept of civilisation—your concept of civilisation—is part of the social structural concept of civilisation but it is not the totality of it. And in that sense I do think that the antinomy of civilisation–barbarity is somewhat built into the general usage of the concepts. And in that light I would like to make a comment on one thing that Evers said, which is that, in the modern world-system, where is the outside environment? But is not that precisely the point? In the world-empires as they emerged historically, the self-definition of what was 'civilisation' was that which was the world-empire, and self-definition of what was barbarity was that which was not. I mean, that is where we originally get the word of barbarity, which means 'those outside the group'. And as the borders of the empire expanded, the borders of civilisation expanded and everything beyond was barbarity. The quality of the modern world-system and what makes it so different in its internal structure from the traditional world-empires is that the antinomy civilisation–barbarity is built into the internal structure. So that xenophobia becomes racism. Racism is really quite different from xenophobia. And I think therefore the first point you made was well-taken, that the expansion of the world-system involved the dual role of the chivalrous knight at home and the violent conquistador abroad. But the second point therefore does not follow, that the mere fact that the social structure expanded to include the whole world destroyed the premises or the possibility of realising its potential as a system. It is quite clearly the mere fact that you have incorporated the process within the system itself is what indeed permits it to expand to include the entire environment of the world. Indeed, in one sense [that fact] condemned it to that.

Evers: Yes, the discussion so far has very much concentrated on this Western civilisation on one hand and what happens or goes along with it. I think that the world-system can be regarded as being sort of integrated on an economic level, but the question that I would like to pose is whether a world-system is possible on a mental level, and I think that creates a good deal of the problems that were also indicated by what Wallerstein argued. As a mental process, world-system is so far, I think, unthinkable and consequently it is permanently on the verge of being differentiated and breaking apart. Economic integration does not mean cultural integration. And the explanation I tried to give—and I think Professor Elias has clarified that point—[is] that the civilisation process always destroys other civilisations. I do not deny that there are civilisation processes going on everywhere, but the civilisation process destroys other civilisations and creates now on the mental level the image of barbarism all the time. And the creation of the image of barbarism is used as a justification for precisely this destruction of other civilisations.

Kaufmann: Well I would question a little bit that thesis insofar as we have excluded until now the issue of political development. If we see that we have since 1948 had the Declaration of the Universal Rights of Man, we see that we had the *Völkerbund* [League of Nations] and we now have the United Nations which includes more or less all parts of the world insofar as they are organised in a modern political way. I think that these are emerging properties which go in the direction of a system formation. Well, one can debate infinitely about the term 'system', but I think that we see in terms of structural evolution that there are new institutional bodies, that there are new visions of humankind which unite all members of a network of societies, to be not too ambitious for the moment, but which has emerging properties to become systemic.

McNeill: May I say a word? I feel a little bit like a fish out of water in this circle because many of the intellectual positions represented around the table are not intimately familiar to me. But I was much struck in listening to Professor Evers's remarks of the omission of two dimensions which, were I trying to discourse on the subject, would figure very centrally. One is religion. The link so far as there are links between court society and the peasants which you spoke of was surely a common description of the Christian religion—however diverse it may have been in its meaning for a courtier and a peasant. And the influence of the church and of the confessional and of the preaching upon behaviour is of course a very distant one—that is, the ideal and the actuality never very closely corresponded. But it is the great institution, and the great ideological flag around which this whole Western society was organised, thought itself to be organised. And this is true of the other great civilisations, the Muslim, the Chinese, the Indian. Each has its own great religious overarching identity. And if you are going to talk of intercultural relationships, it seems to me that to leave out religion is extraordinary, just plain extraordinary. And I must say that's true of the modern world-system as well. It is one of the things I miss in [Elias's] book. And to talk of humankind, without that self-identification—and it is not a dead one: think of Khomeini, think of what has happened in Iran in the last decade.[1] It is not something that is disap-

[1] The overthrow of the Shah of Iran (Mohammad Reza Pahlavi, 1918–80) in the Islamic revolution led by Ayatollah Ruhollah Khomeini (*c.* 1900–89) had occurred early in 1979, and was therefore fairly fresh in the memory at the time of the conference.—eds.

pearing. It seems to me the great competition in the twentieth century is between revived traditional religion in the revolutionary form, whether Muslim or Christian. We are seeing competition between Marxism and Christianity in Latin America today and in parts of Africa. And Islam and Christianity—each is revolutionary faith. Social uprising in the name of these old faiths in competition with the new faith of Marxism. And it seems to me that this is an enormous omission—that I am quite staggered you could make it, that you belong to the Enlightenment in the most extraordinarily naive way.

The second thing is there is something I call technology—that is the changing patterns of communication and transport which govern, affect—govern is perhaps too strong—affect communication from one group to another and the in-group-out-group formation does not seem to me much of ... The idea because there is contact, therefore there shall be assimilation is, it seems to me, quite naive. Contact with foreigners simply reinforces or provides occasion for reinforcing in-group identity frequently. The ambivalence of that situation, borrowing or repelling foreign models, is accentuated by the opening of new communication, but the assumption that the autonomous independence, the cultural, intellectual, moral independence of local groups would be destroyed by such contact seems to me implausible. It may be tempting to take what they can from the foreigner and to strengthen the traditional elements in their society to repel that which they wish to reject. And again, this takes me back to the religious identities and effort on parts of the Muslim world to repudiate Western customs and Western pressures. We see it taking place there transparently and at a high level of intensity. Similar things, it seems to me, are happening in China, India, and in the Soviet Union. Technology alters the intensity of connection and interactions between the different cultural stems of humankind, but that the assumption should be that you get homogenisation—this is the world-system objection. And I am really arguing with you rather than with Mr Evers. It seems to me implausible, to say the least. It is rather that all cultural identities will take new forms in the presence of new exposure to contact with the rest of the world. Or at least this is how I propose to understand it. So, as I say, I feel that there are two dimensions of the world experience. It is not just the modern world which you left out. One, religion, I think is inexcusable. Technology is not inexcusable but it is also a variable that were I trying to cope with, trying to understand the world, I would always want to have in the forefront of my mind as what it is that permits interaction. It is contact, communication, changing patterns of communication that come very rapidly in the last two or three decades, two or three hundred years and the world is indeed staggering under that intensified communication. But that it involves the surrender of local autonomy seems to me implausible, in-group-out-group formation, it is happening within the nation. We are not homogenising, we are differentiating in the United States, in the Soviet Union, in India, around the world. This sort of ethnic tribalism is becoming much (more autonomous), more pronounced rather than ... homogenisation.

Evers: Maybe I am permitted to make a short statement and then we can proceed to Wallerstein's paper. Now I think the critical comments that have been made by Professor Kaufmann who draws attention to the political process of creation of world bodies, United Nations organisations and so on, as well as Professor McNeill's remarks, should be taken very seriously and I think they point exactly at the dilemma we are in when we discuss worldwide processes like civilisation processes or world-systems. Now I feel a bit staggered by this allegation of omission of religion: having formerly spent a few years of my life writing on the sociology of Buddhism, I am surprised to be accused of missing that point. But I think there is something to what you said, and looking through my manuscript there is a short remark about religious revolutionary movements like the resurgence of fundamentalist Islam, but that is about all. Now, in the same way as Professor Kaufmann pointed to political organisations, you point to religions as an integrative or overarching force, and of course, the major religions are called—not without some justification—world religions. But we should not forget that Christianity also in the same way stimulated the rise of Islam. We did not get one world religion, but we got another world religion that constantly challenges, and still challenges, Christianity, as a counter-force. The identity that is still being created by religion leads as much to conflict, as we know, as to an integration of larger units. We are not talking, as you did, about in-group identity that is created by contact with foreigners—we are talking about force, submission, destruction, and that is quite a different matter. And maybe you should not be too naive about what is going on in this world in terms of contacts between the civilised world as we understand it and processes of lower levels of civilisation, or decivilisation, as I tried to identify them. In the question of the homogenisation of the world, I think I have strongly argued, as I understand it, along the lines of Elias. What we get, and I think this is also what you meant towards the end of your remarks, is a constant resurgence of new patterns, of new figurations rather than a sort of homogenised world-system. I leave it to Wallerstein to respond to that, but I think he is also very far from arguing that there is a homogenised world-system—he is arguing that there is precisely a differentiated world-system. I think, however, that these processes of differentiation—whether this is ethnic resurgence or fundamentalist religious resurgence—should be interpreted, not just seen but interpreted, in a theoretical framework. And the theoretical framework that I propose is of dialectical nature, that processes of civilisation always produce counter-movements, and in such ethnicity movements, new religious movements can be interpreted in this way if they are connected with the civilisation process as analysed by Elias in the other more central areas of the world.

McNeill: Oh yes, I did not mean to dissent from the notion that civilisation creates barbarism. I would call it acculturation but it does not matter. The civilising process of acculturation creates the out-group via the in-group. The in-group creates the out-group. It depends on the identity of one group or the other. I thought that was very well said. So I am not—we are not—totally disagreeing. Thank you.

Wallerstein: This previous discussion clarifies anyway what I should do. I was wondering, since I had written a paper which I presume most people have read. I didn't think I should read it. I will try to resumé the main points, but I will do it now in the light of what has been said. I must say that, as a preface to that, just a comment and an anecdote. The comment is, 'Please note that the criticisms of Evers and McNeill that have been made of me go in opposite directions'. That is to say, Evers in effect says, 'Well your world-system, it works if you leave it at the level of economics, but if you take it at the level of culture, it doesn't really work'. Which is the opposite criticism from the one I often get, which is the McNeill criticism: 'You're only dealing with economic variables, basically, maybe political, but you are really leaving out the cultural variables in the analysis'. I don't think the latter is true either. In fact I think I talk very much of religion in *The Modern World-System*, and indeed I think the entire paper that I have written for this conference is on religion. Because I think the religion of the modern world-system is science and the belief in the civilising process. And that is indeed my point.

McNeill: You are a real son of the Enlightenment!

Wallerstein: No, no. I must immediately say that though it is true that I am a true son of the Enlightenment I am also extremely ambivalent about that, and indeed I do not—I underline—myself believe in the theory of progress: I reject it, the theory of inevitable progress which is the theory of the Enlightenment. I merely say that theory is in fact the religion of the modern world-system. And that is indeed the point that I was trying to make in the paper. ... The dual connotation of the word 'civilisation', with which we are all familiar, is not a kind of sociological accident, but is in fact itself reflective of the very nature of the modern world-system as a civilisation.

The Modern World-System as a Civilisation

Immanuel Wallerstein

Et c'est celle instabilité des trajectoires, ce sont ces bifurcations où nous retrouvons les fluctuations de notre activité cérébrale, qui nous sont, aujourd'hui, source d'inspiration.
—Ilya Prigogine and Isabelle Stengers, *La Nouvelle Alliance* (Paris: Gallimard, 1979), pp. 268–9.

The word civilisation, it is notorious, has two rather distinctive meanings, both of which are reflected in the very title of our conference, 'Civilisations and Theories of Civilising Processes'. On the one hand, it is a term with very positive connotations which by its logic is grammatically singular, denoting processes (and their results) which have made people more 'civil'—that is, less 'animal-like' or less 'savage'. When French colonialists in the nineteenth century launched the slogan of *la mission civilatrice*, no one had any doubt about the uniqueness and therefore the universality of the civilisation to which they referred.

I. Wallerstein (✉)
Binghamton, NY, USA

A. Bogner, S. Mennell (eds.), *Civilisations, Civilising Processes and Modernity – A Debate*, Palgrave Studies on Norbert Elias,
https://doi.org/10.1007/978-3-030-80379-7_4

On the other hand, there is the plural usage, in which *a* civilisation refers to a particular concatenation of worldview, customs, structures, and culture (both material culture and high culture) which forms some kind of historical whole and which coexists (if not always simultaneously) with other varieties of this phenomenon. This usage is a little more 'neutral' in tonality than the other is, or rather the ideological overtones are more complex and more subtle.

Lucien Febvre starts his discussion of this very concept of *civilisation* with the assertion, 'It is never a waste of time to write the history of a word'.[1] It turns out, on looking into the matter, that the two meanings have been with us almost from the beginning and they reflect two very fundamental developments in the modern world-system as a civilisation. That is to say, and let me emphasise this from the outset, the very concept of civilisation (a bimodal one) is itself an historical product. But it is the product of a particular civilisation. Does it therefore, by this very fact, call into question one of the usages? Or is it the product of the 'civilising process', thereby calling into question the other?

Before, however, we get lost in this mirror within mirrors, let us review briefly the terminological history. Usage number one, civilisation as the (singular) civilising process, civilisation as the opposite of barbarity, appears first in the middle of the eighteenth century in the works of two quintessential Enlightenment scholars—in French with Mirabeau in 1756, in English with Adam Ferguson in 1767. And Raymond Bloch says,

> the word came at the right moment and corresponded to the notion that had been gradually developing and was destined for a great future, the optimistic and non-theological idea of a continuous progress in the condition of man [*sic*] and society.[2]

[1] Lucien Febvre, 'Civilisation: évolution d'un mot et d'un groupe d'idées', in *Pour une histoire à part entière* (Paris: S.E.V.P.E.N., 1962 [1930]), pp. 481–528, at p. 481. [English translation: '*Civilisation*: evolution of a word and a group of ideas', in Peter Burke (ed.), *A New Kind of History: From the Writings of Febvre* (London: Routledge & Kegan Paul, 1973), pp. 219–57.]

[2] Raymond Bloch (1970) 'Préface' to A. Soboul, *La Civilisation et la Révolution française* (Paris: Arthaud, 1970), pp. 11–13. The *Grand Larousse de la Langue Française* (II, 750) confirms Bloch's view that the first, French usage was by Mirabeau, in the sense of 'action of perfecting' or 'state of high material, intellectual, moral, and artistic evolution'. The *Oxford English Dictionary* s.v. 'Civilization' gives as the earliest English usage Boswell's *Life of Johnson*, XXV, in 1772, in which Boswell recounts Johnson's insistence on using 'civility'

In short, the concept reflected the intellectual triumph of 'rational and experimental science' expressed by the thinkers of the Enlightenment, for all of whom 'civilisation remained primarily an ideal—in very large measure, a moral ideal'.[3]

The French Revolution occurred precisely in the wake of this triumph. This is not the place to review either the causes or the consequences of this major historical moment in our modern world. Two points, however, are necessary to note for our discussion. First, the French Revolution was not only perceived as a fundamental structural change in 'society', a 'revolution'—let us not debate whether or not it really was—but it also legitimated the idea of deliberate, manipulated construction and reconstruction of a social order as nothing had before. This legitimation had an intellectual consequence. If it were possible to *construct* a social world, one was led logically to analyse by careful study the alternatives. The ground was thus prepared for the emergence in the nineteenth century of the social sciences. Secondly, Napoleon's empire first expressed in concrete practice all the ambiguity of a universalising idea with which we are so conversant today. On the one hand, the French Revolution and hence Napoleon's armies were seen not only by the French themselves but by other Europeans as the carrier of a universal idea, precisely that of civilisation, and were welcomed as such. But on the other hand, many of the same welcoming Europeans very soon reacted as local 'nationalists' against French 'imperialism'. Indeed, one could argue that the real birth of nationalism as a political concept can be located in this Napoleonic era.

The ground was therefore ripe for the second usage of the word civilisation, civilisation as a particularity rather than civilisation as universality. The *Grand Larousse*[4] cites Guizot in 1828 in this latter sense, but Febvre gives prior credit[5] to a book by Ballanche written in 1819, which he says first used civilisation in the plural. The two meanings would henceforth remain with us because the antimony universalism/particularism not only became ever more acute but, as it slowly became clear, was inherently unresolvable within the premises and practices of the modern world-system as a civilisation.

whereas he, Boswell, prefers 'civilisation' as preferable 'in the sense opposed to barbarity'.—eds.

[3] Febvre, 'Civilisation', pp. 499, 505.

[4] *Grand Larousse*, II, p. 750.

[5] Febvre, 'Civilisation', p. 507.

The problem is structural. In an historical social system that is built on hierarchy and inequality, which is the case of the capitalist world-economy, universalism as description or ideal or goal can only in the long run be universalism as ideology, fitting well the classical formulation of Marx, that the ruling ideas are the ideas of the ruling class. But if this were all that universalism was, we would not be discussing it today. Universalism is a 'gift' of the powerful to the weak which confronts the latter with a double bind: to refuse the gift is to lose; to accept the gift is to lose. The only plausible reaction of the weak is neither to refuse nor to accept, or both to refuse and to accept—in short, the path of the seemingly irrational zigzags (both cultural and political) of the weak that has characterised most of nineteenth and especially twentieth-century history.[6]

Marxism, the great oppositional *Weltanschauung* of the modern world, solved nothing in this regard. In some ways, it compounded the problem. As Abdel-Malek has noted quite correctly, Marxism is perceived in the 'Three Continents' as 'the most advanced critical synthesis of Western civilisation and cultures '.[7] And therefore what? One response is that of Abdallah Laroui. It is to perceive the existence of two Marxes—Marx the neo-liberal, the 'scientist', and therefore the Western ideologue, and Marx the student of 'backward humanity', the historicist, the scholar. To the former Marx, Laroui says, 'The Third World will always have the reaction that Europe had to Napoleon, the Napoleon of universalisation by force of arms, for this abstract universalisation by imperialism is a disguised murder'. To each his Marxism, he says, the Third World intellectual 'will express his or her own version, born out of the conditions of the world in which they live'.[8]

This problem of the ambiguity of civilisation/civilisations is not merely a political problem, if 'merely' can be used of such a major issue. It is even deeper than that, since opening up this issue reopens the fundamental issue of the very nature of science, which seems to me to have been the

[6] I have tried to analyse this at greater length in my *Historical Capitalism* (London: New Left Books, 1983), chap. 3. See also the brilliant essay by Abdallah Laroui, 'L'intellectuel du Tiers Monde et Marx, ou encore une fois le problème du retard historique', in *Marx and Contemporary Scientific Thought*, Publications of the International Social Science Council, 13 (The Hague and Paris: Mouton, 1969), pp. 266–83.

[7] Anouar Abdel-Malek, 'Marxisme et sociologie des civilisations', in *Marx and Contemporary Scientific Thought*. Publications of the International Social Science Council, 13 (The Hague and Paris: Mouton, 1969), p. 498.

[8] Laroui, 'L'intellectuel', p. 281.

basic building-block of our present civilisational consensus. There is a sense in which this consensus had already been achieved at the dawn of the capitalist world-economy. It was located in the triumph of the so-called Baconian spirit which John Herman Randall defined as follows:

> Not power over men, but power over Nature; and that power is the fruit of knowledge. Nature to be commanded must be obeyed; not by the anticipation of Nature in some magic dream, but by the study and interpretation of Nature will there rise the kingdom of man.[9]

This Baconian/Cartesian/Newtonian worldview posed science against magic, anywhere and everywhere, and always. This worldview logically could tolerate the concept of civilisation only in the singular. Those imbued with the scientific spirit were civilised and civilising; and the others were not. No doubt, many resisted such implications in the name of humanism, or of relativism, or of the concept of original sin, but after five centuries it is hard to contend that science as the 'disenchantment of the world' has not come to predominate our intellectual world, in ways both obvious and overt and ways more hidden and deep-rooted.

The link between science as the 'disenchantment of the world' and capitalism as a civilisation based on the 'rational pursuit of renewed profits' has been stated many times, and no doubt most influentially by Max Weber. It is a pity his many disciples ignore Weber's own warnings about the dangers inherent in this link as 'the rosy blush of [religious asceticism's] laughing heir, the Enlightenment, seems ... to be irretrievably fading'. Weber worries that

> of the last stage of this cultural development, it might well be truly said: 'Specialists without spirit, sensualists without heart; this nullity imagines that it has attained a level of civilisation never before achieved.'[10]

Having said this, Weber typically begs off from pursuing the issue on the grounds that he was now talking in the 'world of judgments of value and of faith'—implicitly not the world of science. Perhaps the moral is to be a specialist with spirit, but only after hours. Freud, that other late

[9] John Herman Randall, *The Making of the Modern Mind* (Boston: Houghton Mifflin, 1940), p. 224.

[10] Max Weber, *The Protestant Ethic and the Spirit of Capitalism* (New York: Charles Scribner's Sons, 1930), p. 182.

nineteenth-century genius who came to fulfil the Enlightenment, was no doubt bolder than Weber in *Civilization and its Discontents*. The whole point of Freud's essay, after all, is to see civilisation as a sacrifice, 'built up on renunciation of instinctual gratifications. ... This "cultural privation" [of powerful instinctual urgencies] dominates the whole field of social relations between human beings'.[11]

While Freud sees this as somehow inevitable in general, its particular manifestation in the modern world was not, is not, inevitable in detail, according to Freud. He raises 'objections ... [to] the ethical standards of the cultural super-ego',[12] since it does not take into account the possibilities to obey, the limits to man's ability to control the id. Indeed, Freud concludes:

> For various reasons, it is very far from my intention to express any opinion concerning the value of human civilisation. I have endeavoured to guard myself against the enthusiastic partiality which believes our civilisation to be the most precious thing we possess or could acquire, and thinks it must inevitably lead us to undreamt-of heights of perfection.[13]

But in the end, he says, 'My courage fails me ... at the thought of rising up as a prophet before my fellow-men, and I bow to their reproach that I have no consolation to offer them'.[14] At least, unlike Weber, he does not seek refuge in the value-neutrality of science, but only in his own human fallibility. Still, one must ask why such keen and wide-ranging intelligences as Weber and Freud, having put their finger on the key dilemma of our civilisation (singular or plural usage), quail at going further. No doubt, the sanctity of the Baconian worldview holds a very strong grip on their imaginations. *Fin de siècle* pessimism as counterpoint no doubt was acceptable, if melancholic, but only if it was self-restrained. Weber and Freud, as we know, gave rise to whole academic industries, but not Nietzsche. And yet today, within the very inner sanctum of physics, this worldview is being challenged, not with melancholic self-restrained pessimism, but with a sober insistence on the range of human possibility. The challenge, raised

[11] Sigmund Freud, *Das Unbehagen in der Kultur* (Vienna: Internationaler Psychoanalytischer Verlag, 1930), p. 63. [English translation by James Strachey, *Civilization and Its Discontents* (London: Hogarth. 1951), p. 90].

[12] ibid., p. 139 [English: p. 910].

[13] ibid., p. 142.

[14] ibid., p. 143

by Ilya Prigogine most dramatically, is precisely against the historic self-image of science, accused of reproducing the very sin it claimed to oppose—the divinisation of the world:

> We echo the complaint that science, and in particular physics, disenchants the world. But it disenchants it precisely because it divinises it, because it denies diversity and natural coming-into-being, which Aristotle made the attribute of the sublimer world, in the name of an incorruptible eternity alone susceptible of being the object of thought in truth. The world of dynamics is a 'divine' world on which time has no impact, from which the birth and death of things are forever excluded.[15]

We are reminded that Kant admired as two inevitable orders the moral law and the eternal movement of the stars, but

> We know that we can no longer guarantee even the stability of the planetary movement. And it is this instability of the trajectories, it is these bifurcations wherein we rediscover the fluctuations of our cerebral activity, which serve us today as our source of inspiration.[16]

That Prigogine is ready to break the idol in ways that Weber and Freud feared to do is less a commentary on his moral qualities than on the evolution of the real world in the 50–75 years that separate their writings. The basic paradigms of social science which were also the basic paradigms of the social movements were products of a nineteenth-century world-system, itself rooted ideologically in a Baconian worldview that had been firmly elaborated and established by the sixteenth century. If these paradigms can be challenged today, it is because the capitalist world-economy as a historical system, what Schumpeter called 'the civilisation of capitalism',[17] is in fact in serious historical crisis, and therefore the most basic consensuses are open for discussion for the first time since they were established. Prigogine echoes a complaint whose most visible expression can be read about every day in our newspapers.

An historical crisis is, however, a major phenomenon, not an everyday event. It involves the coming to an end of one historical system and

[15] Prigogine and Stengers, *Nouvelle Alliance*, p. 266.

[16] ibid., pp. 268–9.

[17] Joseph A. Schumpeter, *Capitalism, Socialism, and Democracy* (London: George Allen & Unwin., 1943), chap. XI.

therefore an historical choice, the bringing into being of one or more successor systems. The choice before us is a double one that hinges on the two usages of 'civilisation'. Will the successor system(s) be more 'civilised' than the present one? What is the relationship of this transition to the existence of multiple 'civilisational' processes, and will the successor system(s) have a place for the concept of multiple civilisations? I will not restate here my views on why the particular historical system which is the capitalist world-economy is in crisis,[18] except to say that the source of the crisis is the cumulative of internal contradictions such that the system will find it impossible, is finding it impossible, to reproduce itself as the same kind of system. These contradictions find their roots in economic processes which lead to solutions of short-run difficulties which create long-run ones and in politico-cultural processes which are the consequence of the increasingly visible strains caused by the economic contradictions, but which in turn create political contradictions.

We are consequently living in a period of transition in my view. But a transition to what? The only reasonable answer to that question is that the answer is uncertain. That in some sense is always true of crisis-periods/transitions. We can never be sure how they will be resolved. But it is particularly true of this transition, because of an historically new element in the picture. If you will allow me provisionally my view that there have only existed three fundamental varieties of historical systems, we can notice something important about the history of their temporal coexistence on the planet Earth. My three varieties of historical systems I call mini-systems, world-empires, and world-economies. The definition of each starts with the empirical existence of a real social division of labour with integrated production processes such that the whole is, from both a material and a social standpoint, de facto autonomous. A mini-system is simply a small-scale division of labour within which one finds a single cultural and a single political process. World-systems then are the opposite in scale, larger divisions of labour within which multiple cultural processes can be seen to operate. Two types of such larger systems can be observed: the world-empire with an overarching political structure, and the world-economy without one.

If we now take the planet Earth over historical time and describe the patterns of coexistence of the three varieties over time, we quickly arrive at

[18] Immanuel Wallerstein, 'Crisis as transition', in S. Amin, G. Arrighi, A.G. Frank and I. Wallerstein, *Enemies of Global Crisis* (New York: Monthly Review, 1982), pp. 4–54.

a periodisation which has only three moments thus far. Moment one is before about 10,000 to 8000 BC. At that time, there were presumably only multiple mini-systems. We do not know how many, how long-lived a given one was, or even very much about how they operated. Our ignorance is quite large.

From that time up to circa AD 1500, the three varieties of historical systems seem to have coexisted. That is to say, we could in theory (if not in practice) construct maps for any point of time therein and physically locate a whole series of historical systems, reaching an uncertain total number. The pattern of course was constantly changing since each instance of each variety of historical system was mortal. Historical systems came into and out of existence.

During this period, the world-empire seemed to be the 'strong form'. By that, I mean that world-empires seemed to expand and contract largely by a logic internal to them. (I forgo here a discussion of the dynamics of this process.) Also, such world-empires as were 'successful' seemed to have a longer life than any mini-system or world-economy, which seemed on the whole more 'fragile', due both to their internal dynamics and their vulnerability to outside attack. The history we have written up to now has primarily been the history of the expansion and contraction of these world empires. Each time a world-empire expanded, it absorbed surrounding mini-systems and world-economies, destroying their autonomous existence. Of course, a world-empire sometimes also conquered part or all of another world-empire, but this was actually comparatively infrequent. When the world-empire contracted, it left a social void in the abandoned zones, within which new mini-systems and world-economies arose.

Somewhere circa 1500 a qualitative change in this pattern occurred. For the first time in human history, an instance of a world-economy survived its 'fragility' and consolidated itself as a capitalist system. This is the modern world-system we all know and in which we live. Why this qualitative change occurred is not relevant to this discussion, though I have explicated my views elsewhere.[19] When this happened, the world-economy suddenly became the 'strong form'. That is, now it was this capitalist world-economy which expanded by virtue of its inner dynamic. In the process of its expansion, it incorporated and destroyed the autonomy of multiple mini-systems and world-empires. By the late nineteenth century,

[19] Immanuel Wallerstein, *The Modern World-system*, vol. I (New York: Academic Press, 1974), chap. 1.

its expansion encompassed the whole globe. Now another qualitative change occurred. For the first time, there existed on planet Earth only one historical system. We are still in that situation today. It is out of this new situation, anticipated in the eighteenth century and realised in the nineteenth, that the ambiguous double meaning of 'civilisation' arose. We can now see to what each meaning structurally referred.

Civilisation in the singular was the ideological construct of the proponents of the new all-encompassing historical system. Civilisation (singular) implied that the exclusivity of this new system, its all-encompassing nature, was both inevitable and desirable. In the language that became widespread, the capitalist world-economy represented progress because it was 'civilising'. It should be noted that this was the argument not only of the advocates of the system, but also of its major intellectual opposition, Marx and the Marxists. For Marx too (at least in one of his guises) argued that capitalism was progress. He merely added that it was only the penultimate and not the ultimate stage of progress. This was no doubt a crucial modification of the analysis, but it still left civilisation as a singular concept.

The concept of civilisations (plural) arose as a defence against the ravages of civilisation (singular). The defence was sometimes given a 'conservative' tone, sometimes a 'radical' tone. It came to the same thing. It was a rejection of the hypothesis that capitalism in its only concrete existing form, a world-economy in fact dominated by the 'West' was morally or politically 'better' than alternative historical systems. The most concrete alternative historical systems to which reference could be made and was made were the world-empires that had been incorporated into and destroyed by the expanding world-economy.

I have previously referred to the double-bind of those opposing the system. To assimilate to civilisation (singular) was to acknowledge previous inferiority and acquire at best second-class citizenship. To reject civilisation (singular) in the name of civilisations (plural) was to risk immolation in archaisms that may not even have had the virtue of being truly traditional and may further have been crippling because non-growing. There was no solution within the terms of the debate as given. This can be easily verified by looking at the personal and collective histories of the various spokesman of the oppressed strata of our world-system as they tried to utilise the multiple weapons of cultural nationalism (*lato sensu*), weapons which sometimes worked very well against the enemy, but which at other times blew up in their faces.

We are thus before a moral and political dilemma of no mean proportions when we talk about the relevance of the concept of civilisation to our current problems in the late twentieth century. It will not do to try to hide the dilemma by abandoning the concept and avoiding the difficult analysis. Rather, let us embrace it as the central issue of our time.

First of all, what are our real choices? Since we now have only one historical system extant, we have I believe three main possibilities. One is that the single system breaks up into multiple historical systems, each again with a separate social division of labour. In short, we could go back in this sense to the pre-1500 situation (or perhaps the pre-8000 BC one). This seems implausible, other than via the route of a nuclear Armageddon, which I do not rule out, but which also seems to me eminently preventable. If we go this route, our present discussion will prove largely irrelevant.

The other possibilities involve the transformation of the present Earth-wide historical system into a different kind of Earth-wide historical system. This seems to me the more likely. Indeed, I believe we are already in the early stages of that process. But in this case, we have the question of what alternative futures exist, which are more likely, which more desirable, and what we can do to make the more desirable more likely. I shall address each query in turn. Which alternative futures exist? In detail, no doubt, the answers are legion. But in basic structure, there are really only two possibilities. We could construct a system that would be—as is our present one and most of our previous ones—hierarchical, inegalitarian, and oppressive. There have of course always been persons who argued that *all* historical systems (or at least all complex ones) had to have these characteristics necessarily. Alternatively, we could construct one that was relatively egalitarian and democratic, fulfilling the slogan of the French Revolution. There have long been advocates of such a system. Others of course have regarded such advocates as Utopian. If Utopian means socially impossible, I do not agree with this assessment, though any arguments I could bring to bear would necessarily be speculative or deductive, and not based on empirical experience.

If, however, you will provisionally grant that we are talking of two realistic long-term alternative historical possibilities, then the question remains: what can we say about the historical probabilities? The theory of progress in both its liberal and Marxist versions has argued the inevitability (if perhaps a slow one) of reaching a democratic, egalitarian reality. I surely reject any concept that this outcome is historically inevitable. I see no evidence for this whatsoever. For one thing, I do not believe that every successive historical system up to now has been more 'progressive' than the prior one. I specifically do not think that the capitalist world-economy has represented progress over its predecessors. At best, it has been no worse. It could be argued it has in fact been much worse.[20]

It follows from this premise that we need to examine structural likelihoods. What are the pressures in one direction or another? The main pressures in favour of a democratic, egalitarian alternative are usually thought to be those deriving from mass consciousness and hence mass will in favour of such an alternative. The creation of such mass consciousness in turn has been thought to be a consequence of technological development and its social consequences: increased education, increased communications, increased mass access to instruments of power, and ergo decreased advantages of the 'elites'. You will note here familiar themes of the Enlightenment and the nineteenth century about the 'civilising process'. Even the Freudian pessimism about 'civilisation and its discontents' has been mitigated by the belief that a particular health technology of psychotherapy (defined broadly) could and would contribute to this civilising process.[21]

Personally, I do not deny any of this. I do believe that these processes all do operate, and do have the suggested consequences, up to a point. The question is whether there are not simultaneously other processes at work, going in an opposite direction. For each simplification that technology has brought (and therefore accessibility to non-experts), have there not been coordinate complexities? In the twentieth century, improved communications have been said to imply Big Brother at least as much as the ability of the working class to organise. Has not the mass access to instruments of power been matched by the high cost and extremely tight access to the most powerful and technologically advanced of such instruments? I could go on, but the point is simple and hard to refute. There is no clear-cut case for a technological road to one system or the other. At the very least, we must say that we cannot be sure whether technology will ever have irrefutable implications for this social choice.

Furthermore, I believe we have missed a major process in our analysis of processes of consciousness. Most of the discussion about consciousness and ideology revolves around a scenario with two stages. In stage A, those

[20] Immanuel Wallerstein, 'Economic theories and historical disparities of development', in Eighth International History Congress (Budapest 1982), J. Kocka and G. Ránki, (eds), *Economic Theory and History*, vol. 1 (Budapest: Akadémiai Kiadó, 1983), pp. 17–26.

[21] Freud, *Civilization and Its Discontents*.

who have real power control the mechanisms that form consciousness. They have cultural 'hegemony', in Gramsci's phrase. In stage B, by some process or other, the majority acquire a true consciousness of their self-interest. The veils are lifted; the old system collapses, or is reformed. There seems to have been relatively little debate about the possibility that the very same people who created the old veils might construct new ones. But surely, this is not all that implausible. Surely, this is one way we can describe much of past history, and indeed is one way we do so describe it.

Furthermore, the most obvious technique for the holders of current power to construct new veils would be to take the lead themselves in destroying the old system in the name of constructing the new. I myself believe this is what happened in the so-called transition from feudalism to capitalism, although I know this is a controversial position.[22] If, however, this is acknowledged to be a structural possibility, then our attention must inevitably shift from the ostensible political foreground of the struggle between anti-systemic movements and the defenders of the existing system to a close look at the history, nature, probable trajectories, and internal struggles of the anti-systemic movements themselves. For surely if a group were to seek to 'change everything so that nothing changes', the place for it to position itself would be within the anti-systemic movements themselves.

Before turning to this issue, let us discuss the third query. Which alternative future is more desirable? In fact, the modern educational system everywhere in the world preaches on the surface the values of a democratic, egalitarian world. It seems almost gratuitous to defend its virtues. And yet, this preaching is done with such clear and obvious smirking that it is in fact necessary to talk on occasion about these moral fundamentals. The arguments for the inevitability of social hierarchy derive either from the irreducibility of human differentials (some people are always more intelligent or more competent than others) and/or from the necessity of coordination of all complex processes, coordination in turn requiring hierarchy. It seems to me the case is weak on both counts. No doubt, there are human differentials, and one might even assume that they contain a biological component resistant to (perhaps even inaccessible to) social determination. But how significant are they? And would they necessarily imply negative consequences for a polity organised in a truly democratic fashion? And even if they have been significant in the past, is not the impact of

[22] Wallerstein, 'Economic theories and historical disparities of development'.

technological and educational advance such as to reduce their significance? Finally, even if there remained some residual impact on collective decision-making, are there not social mechanisms (for example, structuring delays into decision-making processes to leave more time for collective analysis and reflection) that might reduce the impact further? I do not propose to write here the disquisition that would answer these queries; merely to pose them, however, indicates, it seems to me, that the spokesmen for the inevitability of hierarchy based on the irreducibility of human differentials have a difficult case to demonstrate.

There is then the second argument against the possibility of an egalitarian world: complexity requires coordination which implies hierarchy. Here the argument is usually an empirical one. It has always been thus; ergo, it must always be thus. But this kind of assessment seems to me to underplay (indeed to ignore) the incredible amount of human collective ingenuity that exists, also a lesson to be drawn from past history. We have been inventing institutional structures for at least 10,000 years, and the later ones always were unpredicted in the earlier stages. Human sociability is too young as a biological phenomenon for us to announce grandly that complexity can only be coordinated via hierarchy. We know on a small scale that this is not necessarily the case. And the technical difficulties of complex information-gathering, storage, and retrieval are in the process of being enormously simplified these days. Here I make appeal to our knowledge of biological adaptation to assert that it is impossible to rule out that we can create an historical system that is both complex and egalitarian.

Having engaged in this elementary, perhaps simplistic, but nonetheless necessary exercise in arguing the possibility of an egalitarian world, I must then add my moral assertion of its desirability, on the very simple grounds that inequality not only hurts the obvious victims, the oppressed, but hurts as well (if not to an even greater degree) its immediate beneficiaries, by depriving the latter of their human wholeness and their possibilities of self-realisation. Privilege is a barrier to self-fulfilment, forcing its holders into activities they would not otherwise choose and constraining their alternatives. Abraham Lincoln said it quite simply: 'As I would not be a slave, so I would not be a master'. All of what I have written is but a prelude to the important question: if we have a real historical choice before us at this time, and we also have a reason to prefer one alternative to the other, how can we arrive at the preferred social choice? I must start by saying that I cannot give a good answer to that question, nor can any other individual do so. It is not a matter of mere individual insight, but of social praxis socially arrived at. The locus of this social praxis, it seems to me, lies more in the arena of the anti-systemic movements (broadly defined) than in the state mechanisms, the economic arena *stricto* sensu, or the cultural-ideological sphere.

It is a matter of envisaging the overall framework of a transition, which might be posed as a choice between a controlled reconstruction or a looser, less structured disintegration. One of the strongest and perhaps least useful heritages of the Enlightenment is the feeling that since change is possible, it is only possible or optimally possible through rational social planning. We have had planned social change *ad nauseam*, from Jeremy Bentham to the Bolsheviks. And the results have been less than happy. Our rationality has involved rationalisation, both in Weber's sense and in Freud's sense. We have had the worst of each sense: the Iron Cage and the self-deception. Perhaps it is time to experiment seriously with alternatives.

Perhaps we should deconstruct without the erection of structures to deconstruct, which turn out to be structures to continue the old in the guise of the new. Perhaps we should have movements that mobilise and experiment but not movements that seek to operate within the power structures of a world-system they are trying to undo. Perhaps we should tiptoe into an uncertain future, trying merely to remember in which direction we are going. Perhaps we should constantly re-evaluate whether in fact what we are doing is deconstructing an inegalitarian system or reinforcing it. I know this will seem both vague and naive, but our hard-headed organisational biases of the past 100 years have not been all that successful. What is least clear is the role of the multiple civilisations in this process of deconstructing. Surely, we have no more interest in recreating the world-empires of yesteryear than in refurbishing the universalising domination of a capitalist civilisation. These multiple civilisations, however, are indeed the foci of important anti-systemic movements. We may deconstruct more rapidly in their wake than without them. Indeed, can we deconstruct without them? I doubt it. In any case, a new social praxis must clearly be built out of a family of movements that encompasses the wisdom and the interests of all sectors that have been put down and marginalised in our present system. An inclusive family of movements will not only be numerically stronger, it will also have the great advantage of experiential variation and hence of increasing the likelihood of discovering correct paths.

It will not be easy to build such a family of movements, and the opposition will be fierce. Norbert Elias ends his book with a citation from

Holbach: *la civilisation … n'est pas, encore terminée*.[23] Indeed not, and the obstacles seem to me great indeed. But I take heart from this conclusion of Prigogine and Stengers:

> The ways of nature cannot be foreseen with certainty. The element of accident therein cannot be eliminated, and it is far more decisive than Aristotle himself understood. Bifurcating nature is that in which small differences, insignificant fluctuations can, provided they occur in opportune circumstances, spread out through the whole system and bring into being a new mode of functioning.[24]

The circumstances seem to me opportune. Let us encourage by our actions this bifurcating process and let us try to engender a new mode of functioning in which the distinction between civilisation (singular) and civilisations (plural) no longer has a social relevance.

[23] Norbert Elias, *On the Process of Civilisation*, trans. Edmund Jephcott (Dublin: UCD Press, 2012 [Collected Works, vol. 3]), p. 490.

[24] Prigogine and Stengers, *Nouvelle Alliance*, p. 271.

Discussion of Wallerstein's Paper

Transcript

Brands (in the chair): Thank you so much Professor Wallerstein for offering us your grand views combining deterministic and non-deterministic elements and presenting to us the choices we have to confront. Let me open the discussion right away.

Koenigsberger: I have been most impressed by Professor Wallerstein's paper and by and large, I do not want to criticise it at all. I would like to comment on some aspects of it, two in particular. One, the question of whether we can expect further unified world-systems; and the other one, the to my mind very fruitful notion of options which lie open to us and the normative aspect of what he said on this. And I think one ought to stress that he meant when he talked about inequality, that there is a very strong normative …

Wallerstein: Oh yes, absolutely.

Koenigsberger: … aspect in that. Now, my own position is very much between two stools, between what Professor McNeill calls 'microhistory'

Transcript (✉)
Berlin, Germany
e-mail: Artur.Bogner@web.de

A. Bogner, S. Mennell (eds.), *Civilisations, Civilising Processes and Modernity – A Debate*, Palgrave Studies on Norbert Elias,
https://doi.org/10.1007/978-3-030-80379-7_5

and 'macrohistory'. Presumably, I am sitting on the floor somewhere between those. My vision will therefore be rather restricted, but I have the advantage of being able perhaps to make some analogies in both directions—towards the micro and the macro. Now the analogy I want to produce, which I think may well be illuminating and you may find illuminating, sir, is that of the breakup of the mediaeval Catholic church which, by way of analogy—I don't want to go into the details of that or the problematic of that—I would like to call a world-system. Now I have myself written on why this happened and have used for this purpose a model produced by Karl Deutsch.[1] Whether in the company … I am treading on anybody's toes by mentioning Karl Deutsch, I don't know. Happily, I am not a sociologist but an historian, so I can use the sociologists I find useful. Karl Deutsch produced a model of the internationalism of mediaeval Europe which was based on the scarcity of skills and resources. Now that makes a lot of sense. Europe in the Middle Ages wanted a lot of skills and resources which its different parts could not afford. The example which I myself have used is that of the bell-founder who goes to small towns, casts three or four bells, and then obviously he will have to move on, probably a very long way because there would not be any more work for him in that town. And the international situation of the church was based to a large extent on again the scarcity of skills, the skills which the church provided with the advantage of its international language, Latin. Hence, especially in the early history of the church, this moving around of its more skilful persons … This seems to me to be possibly an even more important point than (the one) Professor McNeill has insisted on, communications, which otherwise I would not deny. I think this is also a good concept but I think one ought to air this. Now what happened is that, in the later Middle Ages from the thirteenth century onwards, Europe became sufficiently prosperous. The different regions became sufficiently prosperous that, functionally speaking, the internationalism of the Church was no longer needed. It continued, of course. Institutions as you know—especially old established ones—have a tendency to go on for a long time even when their original functions are no longer being fulfilled. There was the development of internationalism, but more important—in the different large areas of

[1] H. G. Koenigsberger, 'The unity of the Church and the Reformation', *Journal of Interdisciplinary History*, 1: 3 (1971), pp. 407–17; Karl W. Deutsch, 'Medieval unity and the economic conditions for an international civilisation', *Canadian Journal of Economics and Political Science*, X (1944), pp. 18–35.

Europe—it was more possible to develop your vernacular civilisations and within that, if I may use this word, your vernacular churches, churches which depended on the local princes and so on. So that eventually—it was not something which went wrong with European civilisation that this unity broke up the way Toynbee thought—but something which went right. Things went well and therefore the unity was functionally no longer necessary. Hence, the Church broke up.[2] Hence also, all the hopes of Luther, Calvin and the other reformers of substituting their own version of Christianity for that of the Catholic Church were doomed to failure from the start. This was quite regardless of the particular theology or the particular doctrines, and had really only relatively little to do with that. What I would suggest from this is that we are getting to the sort of stage where the universalism that we have known in the nineteenth and twentieth centuries, the world-system of capitalism, is now breaking down. And I would suggest that at least one of the reasons is this Karl Deutsch reason—that the different parts of the world are beginning to function quite nicely on their own. Not that internationalism disappears altogether or that communications disappear. On the contrary, they can continue to grow, but the regional aspect grows faster than the international communications. It grows faster than the upper crust of international skill which is necessary, [and] which are no longer as possible. I would therefore suggest, if one is allowed to shine a torch, a flashlight into the murky future, that at least for the foreseeable future another world-system is most unlikely. I think what we are likely (to) get is the breakup of the world. Now, the other point I wanted to make is that of options. By using the historical analogy, the result of the breakup of the unity of Europe was, of course, in many respects, utterly horrendous. It led to the most ghastly religious wars. That is to say, there had always been wars and civil wars, but now, with everybody having such a good conscience about their own side and regarding the other side as being inspired by the devil, things became quite utterly horrendous. The result of these was not the victory of one or the other or the establishment of a new combined church of all. The result was that a lot of people began to move around, move away from church, move away from the notion that religion and especially theology and the notion of a church should determine life. Now this I have also tried to trace. I have tried to trace it with regard to music: how there was a

[2] For an elaboration of this argument, see H. G. Koenigsberger, 'The unity of the Church'.

psychological shift by people into the music which is largely unideological and, although closely linked with religion, is still, in terms of doctrinal matters, largely neutral.[3] I have also tried to trace it—somewhat more tentatively, but I think still validly—with regard to natural sciences. Natural sciences were also something people could, in a sense, escape to and develop because they were—they had, of course, to establish their own field as against the theologians and this was the well-known struggle between church and science. But more important still, they produced also a new psychological field in which the religious passions no longer impinged. This is very, very clear and quite clearly stated in some of the early papers of the Royal Society. And clearly here are or were very constructive options—options which were not destructive, which did not lead to war and which are the sort of options, if we are to follow your terminology, which I think we ought to look for in our present situation.

Goudsblom: I would like to say something about Professor Wallerstein's presentation, not about substantive parts although that will also come into it, but especially about the conceptual problems which it raises. The theme civilisation or civilisations is, of course, to large extent, a conceptual issue and I would give a slightly different reading of the historical career of the concept of 'civilisation' and draw also therefore different conclusions from that reading about how to use it in our circle today. I think that the concept of 'civilisation' is like the concept of 'culture'. They are really twin concepts and perhaps you would do well to regard them as such as part of one word family. They are modern variations upon a very ancient theme—the ancient theme of the city dwellers versus those who live outside the city gates, the members of the empire versus those who are barbarians beyond the military fringe of the empire. And they always contained oddly the very strong stigmatising, self-congratulatory connotation. And at the same time, there were some empirical elements in it also. I mean we would have had no difficulty in recognising a Greek-Athens city dweller from the peasants from the mountain. Now this very old theme of the more polished city dwellers versus the more rude non-city dwellers, those who live outside the gates, has been given slightly a novel turn I think in the eighteenth century with the concepts of 'civilisation' and 'culture', both of which I think came into being in Western European languages more or less at the same time. Both originally more as verbs than as nouns. One

[3] See H. G. Koenigsberger, 'Republics and Courts in Italian and European culture in the sixteenth and seventeenth centuries', *Past & Present* 83 (1979), pp. 32–56.

would speak of the 'civilisation of manners' or one would speak of the 'culture of the mind', and only later did they develop into independent nouns so to speak. And as such, they differed slightly from the traditional self-congratulatory terms, the 'we'-concepts of city dwellers, empire dwellers in that they had a sort of an oppositional notion, too. They were typically, and Norbert Elias has pointed out in the first chapter of *The Civilizing Process*,[4] words used by bourgeois people who were in opposition against an aristocracy that at that time was still dominant and that, I think, made the bourgeois inclined to emphasise precisely the point you also emphasise, the universality in the concepts of 'culture' and 'civilisation'. It did not have the traditional exclusivity that a word like 'Greek' or 'Roman' would have, or 'citizen', which would always exclude a large part of the world. But 'culture' and 'civilisation' were in a way also, could also be used to include all humanity. And that is indeed what happened, conceptually. When the words 'culture' and 'civilisation' were taken over by philosophers and later on by anthropologists and historians, they were applied in a sense that was meant to get rid of the old derogatory denotations. It did not always succeed. Of course in the nineteenth-century anthropology you had always tension between, on the one hand, the idea that humankind had gone through a development from 'savagery' through 'barbarism' to 'civilisation' and the idea that all of humanity went through 'civilisation'. And there are pages for instance in the book by Tylor, *Primitive Culture*, where you can find on one and the same page the concept of civilisation used in both senses.[5] But anyway the bourgeois idea of culture and of civilisation was carried over into philosophy and into anthropology as a term which could be applied to all of what at that time was still called 'mankind'—we now speak of 'humankind'—that is a slight change in our civilisation. But in a later stage, I think in the early twentieth century, especially in anthropology, but also in history, earlier already I think in history, both concepts, 'civilisation' and 'culture', were also used in the plural, obviously one could speak of the civilisation of the Renaissance in Italy, the culture of the Zuni Indians. That was a next step in the development of the idea of civilisation or culture. In the academic world—it was really an academic conceptual development which has been very fruitful in the early many varied monographs that idea, civilisation and culture,

[4] Norbert Elias, *On the Process of Civilisation: Sociogenetic and Psychogenetic Investigations* (Dublin: UCD Press, 2012 [Collected Works, vol. 3]).

[5] Edward Burnet Tylor, *Primitive Culture* (London: John Murray, 1871).

as something singular that could be put into the plural. And, I think, in the 1930s again within the world of anthropology and sociology—sociologists had joined the movement at that time I think—there was also a growing criticism of the notion that cultures really formed, as Ruth Benedict for example wanted to argue it, closed configurations that … that [put their] stamp upon every individual.[6] But the idea that culture was a formative influence upon individuals remained. And that also could be applied to the word 'civilisation'. …

Bertelli: I think that everybody should agree with Norbert Elias when this morning he said that he understands 'civilisation' as not opposed to barbarianism. And also with Professor Wallerstein when he affirms the existence of plural civilisations and not one only civilisation. But let me add another possibility of interpretation of civilisation, and this is a dialectical understanding of a civilisation inside the plural sub-civilisations. … [One example] is the Italian Renaissance. Well, you know that Burckhardt presented an idea of Italian Renaissance as a lay renaissance, an unChristian renaissance. Nevertheless, [other historians] contended that the Renaissance was a Catholic renaissance, a religious civilisation. … This implies that we can approach this same idea from two different interpretations. Perhaps because inside this civilisation two different minorities express … their own positions. … Today for example when a communist is approved by a capitalist he [sic] must ask himself what he did wrong … he is inside a Western civilisation … Nevertheless he feels himself … near the Eastern world more than … inside the Western world. … That is the same … as it was perhaps for a Calvinist, a Huguenot in Louis XIV's age. Certainly, he was a Frenchman; he lived in France; he understood the culture of his own country. Nevertheless, he felt himself outside this Catholic world. He felt perhaps near the Germanic Protestant world more than … in the Catholic French world of his own age. So, to conclude, I suppose that inside a given civilisation we must discover a dialectic opposition of different sub-civilisations.

Stauth: If I may add, I think according to the problem raised now we should discuss our conceptual frames and I think there is a basic conceptual difference between the type of questions Elias has in mind and the type of questions World-Systems Theory has in mind. So we should look at what really is the difference conceptually between the terms of 'configuration' and 'system'. Your elaboration today and your works … lead us to

[6] Ruth Benedict, *Patterns of Culture* (New York: New American Library, 1934).

say a world-system is an egalitarian distribution of various institutional settings over the world. And let us say that equal types of institutional frames distribute over the world and that the whole notion of system *projects* systematic frames into the individual. That means the whole perspective is a sort of a—well—colonialist perspective, in that there are various individuals from over the world and there is a spread of a sort of 'systematisation' to which these individuals have to stick. I think Elias's perspective … is a totally different perspective. He opposes this very well, as he did this morning in denying this contradiction between 'barbarianism' and 'civilisation'. … I think we should clear up the perspective on the terms 'configuration' and 'system'. …

Dunning: Can I add a rider to this conceptual issue? Norbert Elias talked about the static polarities involved in Professor Evers's earlier paper. It seems to me that there are whole series of these, whole constellations of these implicit in Professor Wallerstein's paper. Determinism–choice, equality–inequality, autonomy–control, universalism–particularism, and there are many more. Now if I can just very briefly illustrate the difference between his approach and the approach that I have learned from Norbert Elias over the past 20 years very briefly. It is an approach that involves not seeing these as mutually exclusive alternatives, because the Western European civilising process as analysed by Elias involves *simultaneously* the growth of inequality via state formation—that is to say the concentration of more and more power in the hands of the expanding state apparatus—and of equality via what he calls 'functional democratisation'. Each of these processes, according to the Eliasian analysis, presupposes the other. They are not mutually exclusive alternatives, which they are made to be if you see them as static polarities, as a simple question of opposition.

Elias: I wanted very much to go in the same direction as my friend Eric Dunning, and I wanted to say, I think, for our future discussions it is of some importance to ask ourselves, are we talking about our ideals or do we try to find out processes which really happened in that way? I think, in what I have to say this afternoon, it will come out very clearly that I try to keep the ideals which I have, like the other men, clearly apart from my intense interest as to how things really happen. I will not go here into the epistemological problem as to how far that is possible. I think that it is possible. But in any case, if I speak of civilising processes, I am concerned with processes which really have happened and, as I think you yourself have said, there is some similarity with Freud, but Freud never really saw civilising processes as long-term processes. So if you raise here the problem as to whether an egalitarian society is desirable and possible, I think it does not belong in this context. We have to, I mean so far as I am concerned, as I say, I want to find out how these processes have really gone on and tell us a great deal yet to be researched and to be found out. I will have to say this afternoon how many gaps there are in our present knowledge because historians are not concerned—except of course, I am happy to say, Professor McNeill—as a rule are not concerned with period transcending processes. Therefore, a vast area has not been researched.

And if I may come … to what my friend Joop Goudsblom has said, the use of the term 'civilisation' in this context is a bit of, how shall I say it, an innovatory problem. Perhaps I should explain to you—I have not explained it very often. It is really that I *discovered in my material* certain changes in the standard of restraint in the seventeenth century. I looked through the manners books and I found suddenly that in different editions the standards of restraints had tightened, had increased. In one textbook, it was said that it was good manners to put one's hands into the common dish. Perhaps ten years later in the same textbook, in a new edition, it was said today people are so sensitive that they resent it if people put their hands into the same pot![7] So when I got through this whole series I had to find a concept which covered this—what I saw there—and I went on to … taking up the familiar concept of a civilising process. I thought that was the nearest term available to cover what I have seen. I know that in that respect, too, one has to distinguish very clearly between the factual use of the term 'civilising process' and the ideological term. In terms of the ideology, you have really, as Professor Evers said, the contrast between people who think they are higher and better and, still worse, the contrast between 'civilised' people and 'barbarian' people. But in terms of a theory of civilising processes, as a social science theory, I think I will try in the afternoon to show that these differences which we express ideologically—the differences between 'civilised' and 'barbarian' are in fact—and there we come to another theme, Professor Wallerstein, which I hope we can discuss—indicative of the fundamental unevenness of the development of humanity. The difference between 'civilised' and 'barbarian' people are merely indicators of the fact that the 'evolutionary'—I would not use this word myself—or developmental process goes on from small groups to larger groups. Why that is so—well, we can discuss. Why there is an unevenness in the development of humanity—that is an interesting problem. But whenever we

[7] Elias, *On the Process of Civilisation*, pp. 91–128.

look, in the fifteenth century BC, or wherever we look, we find unevenness in the development which today is beginning to even out a little. But [these are] problems which one has to investigate by means of theories that can be checked by empirical data. I think that it is a little bit dangerous—[a] distortion of our understanding—if one allows one's ideals to interfere with a real analysis. That is because only then can one act properly—if one knows, if one has a factual orientation. It is only then that one can act in a reality-congruent manner.

Arnason: I have some problems with the typology of historical systems. The three types—mini-system, empire, and world-economy—seems to me far too simple. And I would like to mention briefly three important cases that seem to me very hard to fit into this scheme. First , the system of the Greek city-states roughly from the eighth to the fourth century BC. What sort of system was this? Obviously, neither a world economy nor an empire. To call it a mini-system seems to me very misleading, given the exceptional characteristics of this constellation and the extraordinary social and cultural creativity that resulted from it. Second case, mediaeval Europe. One way of looking at mediaeval Europe is the model that Georges Gurvitch extracted from the work of Marc Bloch.[8] That is to see it as a combination of five different societal types. Let me enumerate them briefly: the most elementary and archaic is the society of lord and peasant. Then there is the international and potentially theocratic society of a church. Thirdly, there is feudal society in the more restricted sense, that is, the society of a nobility superimposed on the relationship between lord and peasant. Fourth, the urban community. Fifth and last, the emerging territorial state and the ethnic identity that develops alongside it. The point for our present purpose is that we have an exceptionally complex combination within a relatively limited territorial area. To lump this together with tribal and archaic societies under a category of 'mini-system' again seems to me quite risky. A third and last case. India. Surely, it makes some sense for the last 2500 years or so to talk about India as an entity, a socio-cultural identity, much too large and complex to be called a 'mini-system'. The point is particularly in contrast with China in that it is particularly difficult to convert this identity into an imperial structure. The most successful empires were imposed from outside—the Mongol and the British one. To draw a tentative general conclusion: What all three cases perhaps have in common is that it is very difficult to explain the systemic characteristics in terms of

[8] Georges Gurvitch, *The Spectrum of Social Time* (Dordrecht: Reidel, 1964).

a division of labour. Why should we limit the categorical historical system to this criterion if we have to take into account all the mechanisms of interdependence, communication, demarcation, identity formation.

Wallerstein: Yes, yes. I am in many different directions so I am not sure in what order I should take this. Let me perhaps take Professor Arnason's point first because in some ways it is fighting me not at the centre, but what I think of it, at the margin of what I said. That is to say, I agree. The typology is too simple. All typologies are too simple and their only utility is that they are simple, that they reduce immense complexity to something simple. In any typology, it is a fact that can always be attacked and you have indeed these three very good cases—at least two of which I have in fact worried about mentally myself as to say what they are. But in passing let me say, none of these are mini-systems—that is not at all what I meant by a mini-system. A mini-system really comes closest to what the anthropologists used to dream of as the primitive tribe which they never had, if I may say in passing, ever, ever observed, because, in order to observe it, it has to be already part of, and it has to be no longer a mini-system, but within a world-empire or a world-economy. That is the condition of its observability. But in fact, none of these are that. Secondly, and you have placed too static a definition on my terms and therefore, the fact is that all structures are constantly evolving. I suppose, not knowing very much about ancient Greece, if I were to give you my suspicion without, which I hesitate to do because I have not really looked into it empirically, I would say that you could probably define it as something which, at some point in time, and I do not know in which century—the eighth or the seventh or the sixth—it had the characteristics of a world-economy as with all previous world-economies and all subsequent ones except the modern world-system. It turned out to be a rather fragile one. It got transformed first into a world-empire. That is what I would call Philip of Macedon etc., and then got absorbed by another world-empire, the Roman, and lost its reality. That would be my guess. As to how you interpret mediaeval Europe I am not sure. I have … only five societal types. First of all, I am not sure what your temporality is—the emerging territorial state is a different temporality from the church and so forth. So, you know, are we talking of the fifth to tenth century, the fifth to the fifteenth century, the tenth to the fifteenth century, the tenth to the thirteenth? All that would have to be specified and I feel it is a very difficult fit. In some ways what one could say is that, one mode of interpreting is that it was an attempt. Mediaeval Europe represented an attempt out of the disintegration of the Roman

Empire to create another world-empire, the Carolingian one, which proceeded to disintegrate and in the long process of disintegration finally became something else. But it is also not very satisfactory. India, I am not sure, however. I would be more categorical. I do not know that it is very useful to think of India as an entity for 2500 years. Precisely because it did not represent in fact a division of labour and I think that is an ideological construct of the nineteenth century to think of India as an entity for 2500 years. There were a series of world-empires that emerged on the Indian sub-continent, which define themselves historically in very different ways. And there were some side by side with each other in various points in time. They absorbed mini-systems. They moved back in space or retracted and new systems grew up within that and yes, if you took it as an entity on the basis of religion, of Hinduism or anything else, it would really be hard pressed. You can impose upon it, which is what I have been accused of doing in another way, you can impose the analyst's unity, but I think to see it from inside is not to see it appropriately as a continuing entity for 2500 years, that is an ideological construct, which does bring me to the Stauth–Dunning–Elias points. Let me say I think there are basic differences here, although I think my position has been mistaken. And therefore, we are not going to get very far with analysing the basic differences unless my position is correctly perceived. Stauth seems to think that I believe that the world-system involves homogenisation of the world. Precisely the opposite. Precisely the opposite. I think the modern world-system is a process of heterogenisation of the world, but I do not believe, and that is the difference, that somehow there are things out there whether they are individuals or group size things which we call states, cultures, or whatever, that somehow just got together in contact with each other subjectively, and reacted one to the other. I do think that structures get created, that the structures then create the sub-groups. That is in fact the nearest thing to a sociological generalisation I am willing to abide by in most of my analyses. Society creates the individual. Individuals do not create society. I do not have a Hobbesian–Lockean view of social structure. And indeed, most of modern social science really does, and, to that extent, world-systems analysis definitely is on the other side. As to the polarities, I thought my point was exactly the opposite. Precisely the opposite. The polarity of universalism-particularism is not a clear-cut choice. The whole ambiguity is that you cannot choose. And that is what is built into the present system. That in fact you are placed before, the only people who can choose are the people who, and even they, I will come back to them and discuss Goudsblom's contribution, even the beneficiaries of the polarity find it an ambiguous choice, but certainly those who lose by it do not know what to do about it. So they are both universalists and particularists at the same time. And in the same speech and in the same thought process. It is exactly the opposite of a clear-cut polarity, it seems to me, that we are dealing with in the modern world-system. Now, Professor Elias wants to distinguish between the factual and the ideological use of the term 'civilisation'. I do not think you can do it. That may be a fundamental difference. That is to say, I do not think you can in general distinguish between the factual content of a concept and its ideological content. All concepts are attempts to capture factual reality. They all have ideological both motivations and implications and the whole game of attempting to pull those strands apart, instead of being willing to deal with them simultaneously, is in fact the religion of science as an ideology. That is exact and that is in that sense that I am not an Enlightenment man at all. You see, I reject that fundamentally. I do not believe that one can do that. I am interested in period transcending processes, although I do not believe in that there are simple statements of that. But that then does put before me the question of whether an egalitarian society is or is not a sociological possibility, because it is surely the case, that is, as I look back over 8000 years I do not find any egalitarian society as an empirical reality. If therefore I am trying to talk about period-transcending processes, I can either deduce from this that since it has empirically never existed it cannot exist or I can try to analyse whether that is an alternative, precisely, in terms of the options, which again are a structural phenomenon for me. Since all historical systems begin and come to an end, and I believe that is built into their structures, all right then, I must ask separate kinds of questions about how they function during the period of their functioning and how the transition process operates. And I suggested that the transition process operates in terms of expanding the range of human option and human choice, at which case I have to pose the question—I cannot evade it—of ideological preferences. And that is exactly it. It is in terms of societal transitions, a transition from one historical system to another, that the ideological question becomes an empirically real question, which it is not at other points in time. It is a veil. It is a veil at other points in time. Of course, there has been a fundamental unevenness of the development of humanity. I could not agree more. And the question again would be—it is another way of putting the same question: Is that built in to the process of human existence? Will there always be a fundamental unevenness of the development of humanity, or is it possible that it becomes relatively more even? You have said that it becomes slightly more uneven in recent years, but one does not know from that whether you think that that is a new trend in human history or a momentary accident. I am asking us to seriously reflect upon that issue—Bertelli, I agree. So let me come to Koenigsberger and Goudsblom because what Koenigsberger said intrigued me and therefore helps me to answer Goudsblom. The analogy. The analogy is a useful one and I am quite sympathetic to thinking in those kinds of analogies which Western history incidentally lends us to the analogy of the modern world. I often talk of a different analogy which is the transformation of Christianity into a world language for a certain part of the world and the transformation of Marxism into a world language. That is going on right now. It is another kind of interesting analogy and how it hides under it all the existing differences. Your description of what happened in mediaeval Europe is correct, but incomplete. Yes, there was internationalism based on scarcity of skills and resources. Yes, because of certain economic and other processes these skills and resources became more dense—let us put it in that way—and therefore one could break up into smaller units without losing immediate survivability and so forth. Therefore, the functional need of the Catholic Church as the institution which both spread these skills and resources and also—what you did not say, but I am sure you agree—protected that bellmaker, his ability to move from area to area—very important, okay? That becomes less important as the role can played by other institutions. So you create a situation in which the unity of the church is no longer as functionally necessary for the operation of people in that world. Other tendencies come along and the resistance to breaking up is less and as you say, cannot then be replaced by another church. Well, and, also as you say, it leads to secularisation, right? People say, 'Well, if theology is less important, so I am interested in music, I am interested in science, I am interested in this', and then the whole phenomenon which we described in literature of modern thought emerges. All of that is absolutely correct. The other element, however, is, whatever is said, you see because simultaneously, what we were doing when we were breaking up with the structure held together by the church, in mediaeval Europe, insofar as it was held together at all, but that was the major single structural thread that holds Europe together and is replaced by other threads that hold it together. I do not know that it breaks up, you see. I would not say that. I would say that first of all there are only two intermediate threads that hold it together. One, I insist, is science. It takes a while, but it substitutes as the ideological thread that holds it together. It is originally called philosophy rather than science. It does not matter. I mean there are other kinds of threads that are holding it together and secondly, although we break up into states, they are unified in an interstate system which is a very concrete and real phenomenon, it seems to me, that grows in strength as time goes on. What you get then is a sense of theoretical equalities—equally sovereign states within the interstate system, equally sovereign individuals, the rise of individualism as an ideology within the state—combined with de facto real inequalities in which the theoretical equality is in fact the ideological basis of permitting the real inequality; and then [comes] the physical expansion of that system to include first ... parts of the western hemisphere later on. And immediately the debate is to whether the people are included. I mean Las Casas in that debate instantly rightly touched there: 'Do Indians have souls?'[9] I mean, shall we say that they know that we are working out this new system in which everybody is *theoretically equal*? does that mean those people too? And the answer has been an ambiguous one because the Las Casas debate is never resolved. On the other hand 'Yes, they have souls' but no, 'but they are children's souls rather than adults' souls', so to speak. They have to be treated differently and exactly then we are doing what I said which was we are bringing the civilisation–barbarity polarity inside the structure instead of keeping it vis-à-vis the structure vis-à-vis the outside. And we are creating a whole new set of language, a way of thinking, which one could demonstrate at many levels. At the level of racism, at the level of sexism. We have not talked about that. It does not seem to me that the relationship of man and woman in the modern world-system is the same as the traditional one. That is to say there are inequalities but they are of a very different kind. And precisely of a different kind in terms of the relationships of the world of work. Because what happens in the modern world as opposed to the previous worlds is, where you had a division of labour you had a recognition that all this work was somehow work and was somehow necessary and what you get in the modern world is a definition of a sphere of work which excludes half the population of the world. That is real work and everything else is somehow not real work, a real devaluation. It is another quite at the level of equality defining inequality. So that then seems to me therefore it is not

[9] Bartolomé de las Casas, at the Vallodolid debate in 1550–1551 about the rights, if any, of the indigenous peoples of the Americas, argued that in the natural order they were free human beings, that they did have souls.

the ancient theme of city versus country, of empire versus barbarian. It is precisely something new.

Yes, first is verbs, then is nouns. That is true. All (old) concepts incidentally. But let me get to your bourgeois versus aristocracy. And the bourgeoisie as themselves now—rebels rather than people on top. I think I go along with that only to a very limited degree. That is the self-image and that is the ideology and, yes, it is indeed that they embrace universalism which is in part a way of saying 'Therefore you know, we are as good as aristocrats and we should have rights and so on'. But that is as far as universalism *as practice* in fact goes, although in theoretical ideology it goes much further. So I am not sure we can forget the form of burden of the word. My whole point is that universalism, that theoretical equalities and so on are, at one and the same time, partial practices; partial practices—and very strong ideologies, which to this day place those who are not the beneficiaries of the reality of the practice—there are some. Twenty per cent of the world's population has equal rights with each other today and perhaps only one per cent had equal rights. But there is still 80 per cent that is out there, and the 80 per cent that is out there, so to speak in one way or another, are faced with this impossible choice which is only one of the many things, not the only thing, that is creating an impossible tension within the world-system. There are many other more important things, because if that were enough, the world-system would have broken down three or four hundred years ago. It is not enough but it adds on to the many other things that I will not talk about now, which create a situation in which we have a particular historical system coming to an end. Now, coming back to your last analogy—Your last analogy was that since Christianity broke down into regional religions—let us put it that way—and nobody could put Humpty-Dumpty together again. So the world-system will break down into—I do not know—ten regional states or something and nobody will be able to put Humpty-Dumpty together again. … The limit of that analogy obviously is that the first thing occurred within the framework of a new system, in my view, which itself was then an expanding system, right? And which had in fact a unity which, one of whose expressions was the theoretical equality of the states and then of course the *cuius regio eius religio*,[10] the theoretical equality of the religions.

[10] 'Whose realm, their religion'—the principle, agreed in the Peace of Augsburg, among the rulers of territories within the Holy Roman Empire in 1555, that the religion of the ruler would determine the religion that prevailed in the territory they ruled.—eds.

But as Max Weber learned when he comes to nineteenth-century Germany, although theoretically Protestantism and Catholicism are equal, in fact, there is an unequal distribution in terms of the social structure between who is Protestant and who is Catholic, which he translated into an ideological problem it seems to me when it was in fact a reflection of a social stratificational problem. So I do not agree with Professor Elias about the agenda. I think we cannot discuss civilising processes unless we recognise where the concept came from: Why suddenly in the seventeenth century your empirical observation, your empirical reality occurs. And why in the eighteenth century people begin to talk about it. In the seventeenth century, it happened. In the eighteenth century, people started to discuss the fact that it happened and we are still discussing it today. Unless we understand the process of the birth of the concept and the concept's role in the system, we cannot correctly analyse the empirical reality of the present as well as of the past. And therefore we cannot in fact reflect upon the major social issue before us which is in fact whether there is a steadily civilising process that has existed over all historical time and will probably continue in a somewhat upward direction indefinitely into the future. I say that is a big question mark and that we have to talk about that, and in order to talk about that, we cannot create a barrier between factual and ideological uses of the term.

The Formation of States and Changes in Restraint

Norbert Elias

There's nothing wrong with controversies. I think and sincerely hope that we shall thrive on controversy. Certainly, what I have to say will be a little controversial.[1] I'm trying to demonstrate that it is possible to do work on long-term processes which is non-ideological and, to the best of my knowledge, reality oriented. That means, of course, that it is fragmentary. Like all research, it is at an early stage. It is, if you want to say, experimental. But maybe a scientific work has to be continued by others taking it up, or be rejected in a controversy. Perhaps I should say if one insists a little in that one exercises self-constraint and does not allow one's wishes and fears, as far as possible, to enter one's research, that is not done because of a *l'art pour l'art* consideration. It is not science for science's sake that one

[1] This is a transcription of a tape-recording of Elias's remarks. It has been lightly edited in accordance with the principles adopted for editing Elias's English in the Collected Works; see Stephen Mennell's 'Note on editorial policy', in *Supplements and Index to the Collected Works* (Dublin: UCD Press, 2014 [Collected Works, vol. 18]), pp. xi–xiv.

N. Elias (✉)
Norbert Elias Foundation, Amsterdam, Netherlands

A. Bogner, S. Mennell (eds.), *Civilisations, Civilising Processes and Modernity – A Debate*, Palgrave Studies on Norbert Elias,
https://doi.org/10.1007/978-3-030-80379-7_6

does it. One does it because the only reliable guides to action are the results of non-ideological scientific research. In a way, we are like doctors. We have to make a diagnosis which is, as nearly as possible, reality adequate. And therefore, a corruption of knowledge, ideological or otherwise, is harmful to the patient.

And we at the moment are very much in a situation—we, threatened by a war—are certainly in a situation where the most realistic form of knowledge is the best which may help to guide us. Certainly, this has a great deal to do with what I wish should happen, namely, that there should not be a war. I think that one of the indispensable things that is necessary in order to bring us in that direction is not merely military disarmament, but ideological disarmament. And that is an additional reason why I think that non-ideological research has in fact the greatest chance to be of practical value, in a world threatened with a terrible disastrous war.[2]

What I have to say I think I can best introduce by taking my cue from one of the finest books Professor McNeill has written. It is certainly not one of the best known of his books, but in my view certainly one of the finest, on the European steppe frontier. In this book he discusses the nomadic infiltrations, which in waves came into Europe again and again prior to the sixteenth century. In fact, it is a staggering picture: if one, as it were, looks at it as a big picture, then one can see how the Huns, how the Hungarians, how the one wave after the other comes until around the sixteenth century the European powers were finally able to stop the inroads being made. He writes here,

> We may perhaps detect a natural ecological cycle in the political history of the Danubian Europe between the eighth–seventh centuries BC when men first fully mastered the arts of steppe nomadry and the sixteenth–seventeenth centuries when firearms, standing armies and the superior elements of modern civilised warfare reversed the old balance between steppe and sown and drove the nomads into permanent retreat. Prior to this reversal in the roles whose more detailed analysis will be the theme of this essay, a pastoral conqueror was likely to celebrate his victory by brutal harassment of

[2] In his last years, Elias was fairly pessimistic about the risk of a 'third great war', and at the time more specifically a nuclear war between the USA and USSR. See, for example, his book *Humana Conditio*, in *The Loneliness of the Dying and Humana Conditio* (Dublin: UCD Press, 2010 [Collected Works, vol. 6]).—eds.

any pre-existing human inhabitants who were so imprudent as to await his coming.[3]

I'm not sure that I would call it an ecological cycle—there is a slight difference in approach between Professor McNeill and myself—I would call it a sociological cycle. But the substance of the quotation is very clear and very graphic. It shows graphically a condition in which not only European countries, but practically all state societies, lived up to the sixteenth–seventeenth centuries. In fact, we have to see it as one structure. The nomads were not there, as it were, by accident. But we have to see it as one structure of former ages: city-states or larger states immersed in a sea of tribal people or nomadic people. These both belong to the same structure—the same figuration, you may say—and again and again the nomadic people ran up against the walled cities or tried to break the big walls which emperors in China as well as in the Roman Empire erected against them. This has to be seen as something which starts from the very time when from the mass of the nomadic people one city area first emerged as a larger form of society—and as a survival unit—with a greater number of people and a higher organisation. In fact, it is symptomatic of what I mentioned this morning, namely the fact that the development of humanity always went on unevenly. Small groups pioneered new forms and it was really those who did not immediately follow the pioneering model, and could not follow it, who then became the assailants who stormed against the more advanced groups or who were dominated by them.

Professor McNeill sees this regularity from the sixth century BC till about the sixteenth to seventeenth century AD. But, in fact, I hope you will agree that this regularity of city-states and territorial states emerging from the nomadic multitude can be traced much earlier. It is from the time when, as far as we know, for the first time the greater organisation of a city-state emerged from the village-state level, from the tribal level. Ever since the tribal people, nomadic people stormed against the city walls and were either defeated and driven back or assimilated, and themselves gradually took over the state rules. That was, roughly speaking, in a continuous setting from the fourth millennium BC onwards, when first the Sumerian towns emerged from the village stage. We can already see wave after wave of nomads coming on, the Akkadians, the Assyrians, the Horites, the

[3] William H. McNeill, *Europe's Steppe Frontier: 1500–1800* (Chicago: University of Chicago Press, 1964), p. 7.—eds.

Hittites—you name it. There are wave after wave. They settle down; it is like a crystallisation process in which at first the small Sumerian towns emerged on the order of magnitude of perhaps Evidu. Old Evidu had about 6000 inhabitants. Then we have the various city-states, on an order of magnitude between 30,000 and 40,000 citizens. And so it was on to the Persian Empire, which had a much larger number of states [inhabitants].

I'm going to take size of survival unit as my cue. Perhaps I have to explain that in greater detail. I mean this growth in the size of a self-ruling or survival unit is not simply a numerical fact. What I wish to say, and wish to say very firmly, is that these figures are not just quantitative figures. They are indicators of a change in organisation—of the transition to a higher, more effective, and usually more productive form of organisation. Perhaps I can use a quotation from one of Keith Hopkins's papers, 'Economic growth and towns in classical Antiquity', in which he observes, in passing, that the very fact of the transition to a higher form of organisation means a higher aggregate production. This is what he says:

> The gross product of the whole Roman Empire significantly exceeded the gross product of the hundreds of tribes and city-states which existed in the same area in the fifth century BC. Settled agriculture, flourishing towns, impressive monuments, the whole panoply of classical culture and of archaeological evidence from north Africa to the north of Britain provide convincing demonstration that a sizeable surplus was being produced and consumed throughout the Roman Empire, and that the average standard of living was higher over a wider area than ever before.[4]

I know that probably Professor Hopkins would not agree with me if I use this as a paradigmatic example for a much more general phenomenon, namely, the phenomenon to which I have just alluded—that if one finds units of a greater number of people, from a certain point on this signifies, as I said, that a different form of organisation, that people have succeeded in integrating more people into a single organisation. In order to do that, in order to integrate more people into an organisation, you have to reshuffle—you have, in fact, to invent new means of constraint and control and, therefore, a different form of organisation.

[4] Keith Hopkins, 'Economic growth and towns in classical Antiquity', in Philip Abrams and E. A. Wrigley (eds), *Towns in Societies: Essays in Economic History and Historical Sociology* (Cambridge: Cambridge University Press, 1978), pp. 35–77.

What we have here—the very graphic description—I would generalise. You have everywhere at first very small survival units or self-ruling units and gradually out of the struggles between these survival units there gradually grow up larger units which, as I say, demand a different form of organisation and a different surplus and so on until—well, we will see what 'until' means.

But let me go back to one of the early stages: for example, see this chart on the blackboard (Table 6.1). You will note how small were the survival groups in which Peking Man lived; probably 25 to 50 people were the average unit then. And we have to think very hard how we can explain that there is a directional process which, over the centuries with all the setbacks, again and again enables human beings to coalesce, to integrate, to organise themselves into larger and larger units. Here you have an example of an unplanned process. Most of you probably know the figures which you see there. It is vaguely known that the population of humankind has grown. But the question of why there should be such a continuous movement in one direction—that question is very rarely asked. Here you have—and this is really one of the reasons why I put it before you—one example of a period-transcending process.

As history is written today, it is usually confined to periods of three or four centuries or more, and historians are experts in short periods of three or four centuries. This limitation has had great advantages. It was the condition under which the enormous mass of evidence, of records, could be

Table 6.1 Elias's blackboard diagram (partial reconstruction)

The size of survival units

Date	*Unit*	*Number of Members*
	Peking Man	25–30
4000–3000 BC	Early Sumerian Towns (Evidu, Lagash)	15,000–30,000
*c.*400 BC	Persian Empire	15 million?
*c.*AD 1	Roman Empire	*c.*60 million
*c.*AD 1	China	*c.*60 million
AD 1700	England	*c.5.2 million*
	France	*c.19.7 million*
*c.*1980	USSR	> 250 [*262*] million
*c.*1980	USA	> 200 [*226*] million
*c.*1980	China	> 1000 [*982*] million

Figures in italics have been inserted by the editors

sifted and be made available to all of us. But I do think that now the time has come—and that is, as I understood it, one of the topics of this conference—the time has come where people will sit together and will say to each other that the traditional form of history writing is only one way to look at the human past. At the moment, the short-term history begins to be a serious blockage for a further synthesis at a higher level. What I am here trying to discuss before you is such a synthesis on a higher level, a period-transcending movement—of which there are many. The next task, it seems to me, is to provide models—testable models—of period-transcending processes which go as this process which I have in mind here, which goes from the very beginning of *Homo sapiens sapiens* to our own time. There's a continuous change, not simply a numerical change but a change from organisations at a lower level of integration and organisation to the next higher level to again the next higher level and again a different level. Let me take the first example, a very well-known example, the example of what Gordon Childe called the urban revolution.[5] By that, he meant the emergence in the valley of the Euphrates and the Tigris, in the delta where they debouch into the gulf, the emergence there from the village-state level of larger units which deserve very well the name of cities. What he perhaps does not see is that these cities were in fact self-ruling states. So from the stage of self-ruling villages there emerged in Sumer self-ruling city-states.

I can briefly tell you what in my opinion is the distinguishing characteristic between a self-ruling village and the self-ruling city-state. The self-ruling village certainly has agencies, decision-making agencies, which take decisions for the whole community, but the people who form these self-ruling agencies are not permanently differentiated for this end; they are the same people who grow their own food on the fields, the same people who act in religious cults; the same people also sit together as a group of 'elders' perhaps and take decisions for the whole village. I myself, in my stay in Africa,[6] have seen many—and that is really where the concept of village-states comes from. I've seen many of these village-states in action, the assembly of the elders, perhaps in many cases surrounded by the rest of the male community [who were] sometimes making remarks to the negotiations of the council of elders. There the same elders are the people

[5] V. Gordon Childe, *Man Makes Himself* (London: Watts, 1936).

[6] Elias served as Professor of Sociology at the University of Ghana in 1962–4, following his retirement from Leicester.—eds.

who grow their own food and who perhaps do religious duties as they all do. Priests are sometimes the first people who are differentiated as separated professionals.

Now, what distinguishes the city-state from the village-state is that in their case the people who form the decision-making bodies are permanently specialised. They no longer grow their own food, but they are permanent specialists in ruling, in one way or another. And they are usually at a slightly later stage. They usually have at their disposal an equally permanently specialised staff of administrators, and above all of military men, because, as I have put in another context, the monopoly of physical force and physical coercion is characteristic of all states from the city-state level on. So there is a very distinct structural difference—a higher form of differentiation. 'Higher' is a word I don't like very much—I usually try to translate it into the 'later' form of differentiation. I like to speak of earlier forms of society, then later forms of society, to take away the stigma of 'higher' and 'lower'. So we have here a later stage of differentiation and integration. The representatives of the integration are permanently specialised for that particular job. It's a form of specialisation like any other. I want to say here that this organisational change, as Keith Hopkins points out with regard to the Roman Empire too, certainly had the making of an organisation which promised a greater productivity, a higher product aggregate and above all a higher ability to protect life and limb of the community against the attacks of the surrounding people who had not made this ascent to the next higher level of integration.

That in fact is today very often [overlooked]. The Gordon Childe model is really rather simple. Gordon Childe, whose great merit is that he first discovered—and perhaps if he had not been a Marxist he would not have discovered—that the productivity, the increase in food supply, which occurred in the Tigris and Euphrates valley was, as he saw it, one of the conditions for the growth of towns. That is roughly his model. It is rich soil, more food supply, growth of population, population organising itself into a higher form, into a later form of organisation, namely the city-state.

I think I would put up a counter-model, and I think many people, many archaeologists today, have themselves made this step. They have said that it was precisely because of the lacking security—physical security—that people gave themselves into the protection of a temple, and at first priests and later specialised kings were the specialists who protected the group of population from physical annihilation by organising special warlike units and, at a later stage, by surrounding the settlement with walls.

Now perhaps I should say that the process from the village-state to the city-state is a process which we must estimate as [lasting] at least 1500, perhaps 2000, years. It's not a sudden change. And we know that the first central figures of this process were priests. I am not able in this talk, simply for reasons of time, to tell you of one of the most fundamental aspects of these early transitions to a higher organisation, but I will at least mention it. I think the earliest problem we have is—and I have discussed it in my paper about the retreat of the sociologists to the present[7]—the reason why actually in former days merchants and businessmen were not the ruling classes. But we have to ask ourselves what structural characteristics of societies are revealed by the fact that all the earlier power holders were either priests or war leaders in the form of kings. What does it tell us about the earlier stages of societies? If there were merchants, they were usually dependent on kings and on priests, if for no other reason simply of physical security, because you cannot trade and make contracts and other things if there is not a war leader or a commander of troops or police forces who guarantees the security which trade on a larger scale demands.

So that is the basic issue: that we have at first a crystallising process, in the course of which, from shall we say a village-state where impermanent 'elders'—which means the family heads—probably ruled, we gradually find a differentiation of functions where at first priests and then after a struggle between priests and kings, the kings became the effective exercisers of constraint. It was, of course, one of the fundamental problems of this whole change, of this whole state-formation process of that level. People who gathered together in a city were now to a greater extent dependent on the strongmen, the priests, or the king, and it was he who now organised them in such a way that the fields could be more productive, that the irrigation channels could be maintained. He was the man or they were the men who protected the irrigation fields and channels outside the city against the invaders who came from time to time and who, as I said, are a structural feature of this thing, and without the physical protection by an army simply, purely as Gordon Childe put it—he gives a rather idyllic picture—you grow more food, the population grows and then you form cities. But the reality is far less idyllic. The reality is that

[7] Elias, Norbert, 'The retreat of sociologists into the present', in *Essays III: On Sociology and the Humanities* (Dublin: UCD Press, 2009), pp. 107–26 [first published as 'Über den Rückzug der Soziologen auf den Gegenwart, *Kölner Zeitschrift für Soziologie und Sozialpsychologie* 35: 1 (1983), pp. 29–40.]

there is a constant struggle, a physical struggle between survival units. And it is this physical struggle—the struggle for survival on a new level, on a non-biological level—which provides one of the main driving forces, if not the main driving force for what we see here. [That is,] smaller units fighting with each other as a result of their survival struggle, [out of which] a smaller number of larger units emerges—from whose survival struggle [in turn] an [even] smaller number of units with even larger [territory and population] emerges, and so forth.

Incidentally, I have set out this model clearly in the second volume of my book on the civilising process under the name of a monopoly mechanism.[8] But although it is, as I believe, very fruitful, although without this model one cannot explain the change in figures—population figures—which I have shown there, one is so little used to get at competition dynamics or to look at the interstate level as the principal driving level that this model has not yet been taken up. I'm sure it will be in the future.

Now, I fear there is too much to say, but I do not want to forget one thing. One thing which I tried to make clear is that there is also an initial struggle between priests and kings, which is one of the standing features of early societies up to our own time. You find in antiquity again and again priests and kings in competition. You find that the Sumerian towns are grouped around both a temple and a palace. You find in Sumer that gradually the priests, the temple, which are the main enterprises, are really the first large-scale economic enterprises that we know. There is no accident that writing is invented in Sumer, because these temple organisations comprising sometimes as many as 1200 people had to have some means of control. And writing in these early days was really simply a means of control for outgoing and ingoing goods in the temple stores and later also in the stores of the kingly palace. But you have this competition between kings and priests going on up to our own time. You can see it in the struggle between the Emperor and the Pope in the Middle Ages. You can see it in the struggle—well, I need not remind you, between Khomeini and the Shah.[9] It's again and again this struggle and it is a very revealing picture because it is probably also invisible. Some of the nomadic tribes,

[8] Norbert Elias, *On the Process of Civilisation* (Dublin: University College Dublin Press [Collected Works, vol. 3].), pp. 301–11. [Original publication, *Über den Prozess der Zivilisation*, 1939; earlier English editions published under the title *The Civilizing Process.*]

[9] See note 1, p. 27 above.

the Indo-Aryan tribes for instance, those who became Hindus, in their setting the priests gained the better of the warriors. In the Hindu caste system, the priest takes pride of precedence of all others, and the warriors are the second-rank caste. In the tribes which migrated to Greece, to Hellas, you find that the warriors are the first-ranking people; and it is quite possible that if the struggle, the conflict between priests and warriors, had taken another turn—had had another result in Greece, among the Greek tribes—we would not have had the flowering of the first philosophy and science. Because that was the first secularisation move, which would hardly have been possible in India as long as the priest groups had the better over the warriors.

I do not want to enlarge this branch of my long-term picture, but I want to show you that certain very lively conclusions can be drawn from a long-term picture of this kind. One more point needs to be made, and that is if you look at these early societies, our conceptual apparatus, the conceptual apparatus which is so much a matter of course to us and seems so absolute in our terminology—namely, the division of activities or spheres into political, economic, cultural, religious and other spheres, particularly the division between political sphere and economic sphere is hardly applicable—is conceptually distorting—if you apply it to early state societies, such as those of Sumer. It's still difficult to apply it even to the Roman Empire because if you have societies where the main source of riches is war-like booty, if those who can lead successful wars are likely to be the richest in the land and at the same time the most powerful in the land because they command the army, I think the separation between the economic and the political is just not possible. There is a very nice saying, a Sumerian proverb which runs: 'It accumulates but does not suffice. It is expended without pause. Upon the property of the king lift not your eyes.' That is to say, the king has so much to pay out. You must understand that he gets all the riches. Don't lift your eyes against them. What we call the class struggle in our societies against entrepreneurs was there, if at all, directed against the king as the richest. And for a very long time we find that merchants and traders are, as it were, private people only whenever the government of the kings get weak. That is also true of the priests. As a rule, we find an alliance between priests and kings in which the priests play a somewhat subordinate part, but are together with the king upholding the rule. That is the same solution we find in later states too.

Now, you find this setting: a powerful city as the core and an exploited rural countryside as the area on whose surplus the town lived well; you find this pattern pretty well till the end of the Roman Empire. In fact, Rome in its early stages till the late empire is still such a conquering state which, driven by the competition with others—with Carthage, with the [Mithridatians, Sarmatians?] and others—accumulates a hinterland, on which it can live, from which it can [gain] rich booty, from which its civil servants can get tribute, from which its civil servants can, as it were, rob and bribe as much as they can. That you cannot quite understand unless one sees that Rome was an island set in an ocean of tribal people. They tried by means of a Roman wall to keep the tribal people out and prevent them from threatening their own towns but, for reasons into which I don't want to go, they did not succeed.

Let me say, in this slow progress through time which I am making, let me say at once that for me perhaps one of the most useful though terrible examples is the running down of the rest of the Roman Empire. Just as you can see the growth of the state organisation as you go from Sumer to Akkad, to Babylon, to Persia, to Alexander and you see how they experiment with new means, always to have a larger and larger organisation to control. Just as you see this building up of the large-scale organisation, which is really a learning process, a process in which rulers learn how to control larger and larger groups of people. Now it's the opposite in the running down of the Roman Empire. You can see there, if I can put it in a nutshell, that a high state organisation is based on the functioning of the two central monopolies of physical force and taxation. And you can see very clearly how gradually both are running down. Less taxes are coming in; as a result, it becomes more difficult to maintain an army; and, because you cannot maintain a large army, fewer and fewer taxes come in. And then you can see how gradually out of the money economy, which presupposes a large state-society, there comes a return to the natural economy. The taxes begin to be levied in the east of the empire, and in the western Roman Empire as well, in wheat and other things. And that is used by the kings, by the emperors in order to feed the army; and then you can see how gradually the people from the cities were stormed by nomads—often three or four times in a year. Trier, I think, was stormed four times in one year. You can see how it runs down. The people abandon the towns. They go into the countryside. And the monopoly of physical force gradually disappears; and you find what you would expect to find if you make an experiment. Namely, the strongmen who are large estate owners organise their private armies and they rule. Now as the central authority no longer can protect them, they rule by protecting the peasants against marauders, but also by exacting as much from the peasants as they can. No doubt, states are exploiter organisations in the initial stages, and for a long time they take ruthlessly what they can by means of taxes.

And here I should make a general remark about constraints. It is a moot point whether humanity could have attained the level of self-constraint which we represent today without going through long periods of external constraint. If I put survival units—such as villages or states—in the centre of my picture, it is not because I am a lover of states. I am not. The state is a compulsory organisation and I personally don't like compulsion from outside very much. But I don't think I am allowed to have—or can allow myself—personal feelings in these matters to guide what I can see in history. And what I can see in the human past is this: that in fact human beings learnt to act on insight and self-control and with kindliness only through long periods of external constraint. You cannot expect human beings to do that, as it were, from the beginning. This is one of the long processes in which they learnt to live with each other in a reasonable way. That, we have learnt—was certainly not there from the beginning. And certainly survival units and states were one of the main real forms of grouping in which people began to learn, but which of course were enormously cruel and were at first administered entirely for the sake—and quite ruthlessly for the sake—of those who had the power.

So, if you look at the Roman Empire and its downfall you have a very good example as to how the organisation of constraint which the Empire represented gradually gives up its constraint function, and can no longer exercise it. And then it happens as we would expect it to happen as I say, [there forms] the pre-form of feudalism, the local strongmen who in later times were also armoured knights, sitting on their horse.[10] At that time, they were probably still foot soldiers, and they exercised the real power of constraint.

Now I do not want to go here into the whole process and the reasons why European countries then succeeded in a new integration process, in a new state-formation process. Part of it I have set out in the second volume

[10] Elias is alluding to the fact that in many European languages, the word for 'knight'—such as *Ritter* in German, *chevalier* in French, *caballero* in Spanish, *cavaliere* in Italian, *riddare* in Swedish—means 'rider' or 'horseman'.—eds.

of my book *On the Process of Civilisation*, but there is much more to be said than I could say today. The main point which I want to make here is that you see the breaking up of a large unit into smaller units and then an unplanned process in the course of which [there is] a reintegration partly through, or in connection with, the existence of the Christian church—so first into smaller and then into larger and larger centralised units.

I personally do not think that this new reintegration process could have happened without antiquity. I think in fact one of the most important aspects, which I can here only mention in passing, is the continuity of the knowledge tradition, which surpasses various periods, which runs from Mesopotamia over the later antiquity, the Greco-Roman antiquity, to the Middle Ages and to our time. I think one can show very well, and with great exactness, that this is one great tradition of knowledge transmission and without this immense knowledge transmission, this whole process could not have taken place. Here you have another of my heresies. I think that the development of knowledge is just as fundamental a process, as basic a process, as the economic development or as the development of the physical force in the form of states. So I mention here in passing that the tradition of the Latin language which goes through the Christian churches to our time, which transmits knowledge to our time, is a very important one.

Perhaps you will allow me to give at least one example. Development: what I am discussing here is in fact not history, but a development. I think that in the near future, one will see that the historical relativism which puts one epoch on the same level as every other is not tenable. There are differences of levels in the development of the past, and so one will have to replace, or rather to supplement, the concept of the human past in the form of history by the concept of the human past as a development—no longer in any speculative manner but as a purely theoretical concept which can be verified by the evidence. You see for instance that the church of the Middle Ages certainly was a going back to a pre-secular, pre-scientific stage. But it was not a going back to the same level which existed before science came up in Greece and Rome. It was a religious, priest-dominated, kind of knowledge on a new level. So, for instance, concepts such as theology would not have been possible without the preceding concept of philosophy. Concepts such as cause and effect, which was taken over by the church theology, were not existing in the mythological cult knowledge prior to the Greek science. So if one speaks of a sequence of stages, as one must do, it does not mean, as usually supposed, a going back to a previous stage. Also, feudalism was not a going back to the tribal level because here, too, you have a new stage, a stage which was going back from the centralised stage but it was not a going back to the earlier stage as it was before.

So let me just make a big jump and remind you of the fact that you can see the processes of integration of smaller units into larger units very well in our own time. And there you can feel the enormous force which the mechanism of competition has in that respect. You see it in Africa. You might very well ask, 'Why do not the African groups go back to the pre-colonial tribal level which they had before? Why do they now integrate on the state level? Why do we live today for the first time in an age where all tribes, all units on the tribal level, have practically disappeared, or are so powerless they don't count any longer?' You can see that the state-formation process goes on in Africa and you can see the same process going on all over the world.

You see it, for instance, in Europe. The European countries are under very great pressure at the moment to integrate into larger units. And you can predict—if you want to say what options there are. There are the options that the European nations integrate around the two superpowers, so that the eastern European states coagulate and integrate as a Russian bloc, as a larger Soviet empire, and the western states concentrate, become part of an American empire. That is one possibility of integration at the higher level. You can take as an option that they integrate among themselves into a larger unit of a European federation of states or they can remain fragmented as they are today. Those are the three options which they have before them, but I don't think that there are many more.

And you have the same process now going on at an even higher level. You can see that there are pre-forms of integration of all states. The question there too is 'What options are there?' There are the options that one great power may attain hegemony over the whole world and then integrate the rest of the world—that it, the world, becomes a Russian, Chinese, or American empire. That is one option. The other option is that we remain for a time fragmented with a danger of wars as we are today. And the third option—still rather utopian, but to me at least very desirable—is the option of a world federation of states in which all violence is excluded between states. Because that, of course, is what we are aiming at. We are aiming certainly at a condition where the old tradition, that conflicts between states can be worked out by means of violence, should give way to a condition in which people go to a law court and give their quarrels and conflicts to a court of law. For that, one has to establish a unified world federation of states. I think it very unlikely that any one of us shall see that in our lifetime. But it is the option.

All this development, of new state organisations on a different level, has very much to do with civilising processes. Because the external constraints structure, which is represented best through the various levels of state formation, expresses itself in various forms of self-constraint. I've actually described in my book that the decisive thing about a civilising process—or one of the decisive things—is the extent to which the external constraints are arranged in such a way that they can transform themselves into self-constraints. That one can easily exemplify. If a child is beaten every day to do what it should do, it never learns self-constraint. It learns only to beat others in order to follow its whims. In order to produce self-constraint you have yourself to be very much more civilised and to exercise constraint yourself—as a fair parent, as well as a ruler. But that may be enough. I want to say this is what we are aiming at. We are aiming at greater civilisation, which means to me acting more on insight, self-constraint, on regard for others, on kindliness, than on external constraint and compulsion of others. We have to see what external conditions are necessary to achieve this aim and I think, though it will not be realised in our time, that the federation of states which excludes war from our doorsteps, and a gradual ideological disarmament of the great antagonisms of our time is one of the necessary steps in that direction.

Discussion of Elias's Paper

Transcript

McNeill: There were one or two more or less trivial matters that I thought worth remarking on. The first is to say I do not think you are correct in saying that in the fourth millennium you have this interaction between the early cities and the barbarian world that you describe. I do not think it is as early as that. The beginnings of barbarian encroachment upon the Sumerian landscape I do not think you can date until the second half of the third millennium. In the beginning, I think the barbarians were not organised sufficiently, militarily, to constitute a serious rival to the emerging city states. Now that is just trivial but it is still perhaps worth saying. The second is a little more insignificant. Your example of the Roman Empire as a sort of city state, writ big seems to me not a typical case. That is, the world-empires that Mr Wallerstein described …

Elias: As an exploiting state.

McNeill: I quite agree that the states were exploiters of the countryside, but the city state core of the Roman Empire is really different, as the Chinese Empire and the Persian Empire are not city state empires.

Elias: That I never said.

Transcript (✉)
Berlin, Germany
e-mail: Artur.Bogner@web.de

A. Bogner, S. Mennell (eds.), *Civilisations, Civilising Processes and Modernity – A Debate*, Palgrave Studies on Norbert Elias,
https://doi.org/10.1007/978-3-030-80379-7_7

McNeill: No, the example you gave was of the Roman Empire as a city-state writ large and this is atypical of the large states of antiquity. It is a small point.

Elias: May I say, I mean, is it not equally true of Babylon?

McNeill: No, it was not equally true.

Elias: There was a big city in the Babylonian Empire?

McNeill: There was, indeed. But the power system behind Hammurabi is not based upon a city state in the same way that the Roman Empire is based upon a city state, I think. Do not take me as an expert on ancient Babylonian history. I'd be much firmer with the Romans, with the Persian Empire, with the Chinese Empires, which are clearly—the Chinese Empire is clearly not a city state.

Elias: I never said that.

McNeill: No. No. The real thing that I wanted to raise with you, the thing I think is important is your remark that self-control is only possible after long periods of external restraint. It seems to me—this is where I found difficulty with your book when I read it—with great interest, but also with some reservation—it seems to me that in what I will call simple societies there is a very high level of self-restraint.

Elias: If I may interrupt. Allow me to correct myself. I have used a too sharp formulation, really against my own conviction. I think self-control is a human universal. I mean there is no society without self-control. This is really my bent. All-round form of self control—that is what I should have said. That develops rather late, but there is no society, you are absolutely right to raise this point and I give you the point. You are right. There is no society without self-control.

McNeill: OK, and in what I call simple societies the level of self-control seems to be very high and external control relatively unnecessary, relatively weak.

Elias: It is usually patchy.

McNeill: Yes, and then you get the situation of early—a civilised situation which people with different self-control systems or different cultural systems are thrown together in which there is no over-arching agreement, no consensus. And this is the point at which the power system, the political, external control becomes important. And it is only when strangers are meeting one another that an external control system becomes necessary. And it is a slow process indeed by which self-control grows up within that external control system. Now the picture you gave of court society was master, really master.[1] A beautiful case of what happened in early modern Europe. But it did not extend to the whole of that society. There were lots of people who did not share that court society and for whom the imitation of court society was not particularly important—even positively repellent. They did not want to become courtiers. There was a certain disdain for the falsity, the untruthfulness of courtiers, etc. So if the radius of the figuration is a limited one, however powerful it is for the several hundred people—I do not know—several thousand people perhaps, who actually took part in it. And so the development of self-control that internalises a manner, pattern of ways, civilised behaviour seems to me very problematic. We have these large control systems—the state, the compulsions of the state, now, that extend beyond the circle of voluntary consent in most cases in modern times as in ancient times. It is when strangers with different internalised moral codes meet one another that the external constraint becomes necessary. And in the modern city—at least in the United States—you see this absolutely transparently. The different moral codes, the different social groups in our communities are in critical disjunction and this is where the police and the armed force becomes the arbiter. Not always very comfortable. So this is my ...

Elias: The last point has not become quite clear to me.

McNeill: Let me say it again. It is when strangers meet, strangers with differently internalised norms of behaviour, of self-control, that an external police force becomes a necessity and is invented and imposed, otherwise you have standing civil war.

Elias: Do you mean without, in other cases the self-constraint is there without police force?

McNeill: Let me use a piece of language which I happen to stay away from. Primary societies—this is an old sociological word are, I think, normally held together by internalised norms without that external compulsion figuring very prominently. That is not true.

Elias: Thank you very much. I am very grateful for that contribution and that you allowed me to correct myself with regard to self-control in general. There is no society with a zero point of self-control.

Koenigsberger: You issue a kind of challenge to historians about your idea of the 'monopoly mechanism' which you discuss in the second

[1] Norbert Elias, *The Court Society* (Dublin: UCD Press, 2006 [Collected Works, vol. 2]).

volume of your book[2]—a challenge that has not been taken up by historians. May I comment on this particular point? Now, the example which you yourself used in your second volume is that of France after the collapse of the Carolingian Empire and the gradual bringing together of the increasing size of units and the bringing together of the new kingdom of France. But it seems to me in the first place, it takes very long, in the second place, the units which eventually emerged are entirely arbitrary.

Elias: You mean the state units?

Koenigsberger: The state units which occur.

Elias: I did not say anything …

Koenigsberger: But the point is while they undoubtedly had a tendency towards monopoly it seems to me that it is very easy for other forces to intervene to prevent this tendency from working itself out. And it has not worked itself out. Western Europe remained divided. In the end, you were forced to produce the neo-Rankean view[3]—and Professor Wallerstein said this too—to extend the whole area of monopoly competition to the whole world. And in the end, the kind of competition which now appears seems to be so different that I do not think one can really compare it with what happened in the thirteenth-fourteenth-fifteenth centuries.

Elias: If I may say so this is a weakness of short-term historians who do not see that there are similarities between different periods.

Koenigsberger: Yes, but the similarities have to be such that there will be some kind of *Gesetzmäßigkeit* [law-like quality] that they will lead to some result which one can foretell within a reasonable future and this I simply do not see. May I give another example of this and take the Middle East? Now, it seems to me that if you take the whole Middle East ranging from Egypt through Mesopotamia to Persia and back to Macedonia and Byzantium, you have a constant change of centres of empires. They change from Babylon to Persia to Macedonia to, back to Persia, to Egypt now in terms of the Islamic conquest; then again to Neo-Byzantium and the Muslim conquest. Now, what happens in each case is that, you have for a relatively short time—quite a short time in your terms—short enough for me, as an ordinary historian, to grasp, 200 years, you know. You have the

[2] Norbert Elias, *On the Process of Civilisation* (Dublin: UCD Press, 2012 [Collected Works, vol. 3]), pp. 301–11.

[3] This is an allusion to the nineteenth-century German historian Leopold von Ranke (1795–1886), often regarded as the founder of the modern academic discipline of history. He formulated the idea of a modern European *system* of states in *Geschichten der romanischen und germanischen Völker* (Leipzig: G. Reimer, 1824).— eds.

establishment of an empire centred on one of the units. Not an integrated empire at all, but an empire specifically run and centred on one unit, not a city state, but a much more tight state. And all the parts in between—Syria, Egypt, Muslim Mesopotamia—are quite happy to be a member of the Persian Empire, of the Byzantine Empire, of the Islamic Empire from Egypt, of the Islamic Empire from Constantinople. They hardly ever defend themselves. They hardly ever worry. They do not even consider themselves to be part of this big empire. I do not really see where your monopoly law comes into this at all. I do not think it works.

Elias: You have not said anything to describe it as you just have pointed out to the various differences but you never ask yourself what all states have in common.

Koenigsberger: Oh, yes, I did. I said they are empires based on one relatively small organised state in the area and because, presumably because—speaking in your terms—because of this and because they fail to integrate the whole empire they remain very ephemeral. They never last more than two–three hundred years.

Elias: Now you make a long-term model.

Koenigsberger: Yes.

Elias: Which you really should not [as an historian[4]].

Koenigsberger: No, I am perfectly willing to make long-term models if they work. I do not think that the 'monopoly mechanism' is more than a tendency. It is not a complete model.

Elias: Well, you will have to ask yourself, I think, how you explain that this regularity, smaller units competing with each other being driven to form larger units or remaining smaller units and struggling, which you see again and again as I say in Africa, in Europe today. The persistence of this fact you will have to explain. I mean, you cannot simply get to rest on having one period before oneself. If there are similarities such as these, smaller units coagulating into larger, and you see in antiquity really very clearly—and I think Professor Hopkins also has made a remark to that effect—how empires, how kingdoms grow and grow and grow and each time you have to ask yourself, how do they manage? What means of constraints do they have to keep a larger empire together? So all that are questions which surpass the present non-theoretical approach of the historians. You cannot do these things without a theoretical framework.

[4] The transcription did not catch the end of Elias's remark, but this was the gift, as we remember it, of the humorous jibe he made to his friend Koenigsberger.— eds.

Stauth: I have to comment concerning Oriental societies. As an example, you yourself pointed out very clearly how the monopolising mechanism came into effect through the wall-building against outside tribes, against eastern Europe, as you just elaborated here. Now, if you put this back to the type of setting you were referring to concerning Oriental realms, it is quite a different situation where you find that tribes move into the centre of the realm in itself in constructing hierarchies and never been sort of externalised by the given structures. And the best illustration of this type of processes, which would be nice to discuss in relation to the problem of self-constraint, is the civilisation theory of Ibn Khaldun, in exactly this context where it is always a type of a circle of re-establishing realms through the establishing of tribes as a sort of a bigger solidarity group, a group of stronger solidarity as a new hierarchy in the centre of the realm. So it is not that solidarity in that sense of self-constraint comes as a result of external forces, but solidarity in this type of theory comes out of sort of a primitive setting, through its stronger coherence, making itself the ruling mechanism—as you said yourself—in a quite open setting of a realm. So I think in terms of an Oriental setting, and a Western setting that we have to clearly separate. And your question was 'Where is the monopolising mechanism there?'—I would say it is a different monopolising mechanism.

Koenigsberger: Are you allowed to make a difference between Oriental and Western in terms of Professor Elias's theory?

Evers (in the chair): May we leave this to Professor Elias to decide?

Elias: I think I have to read Ibn Khaldun. Has it been translated into English?

Stauth: It is translated into English—not the whole. *The Introduction* is translated.[5]

Elias: Thank you very much. I shall certainly try to read it.

Bogner: I wanted to comment on the criticisms made by William H. McNeill on the model of development of constraints, of self-constraints, as it has been put forward in the second volume of *The Civilizing Process*—by the way more than 40 years ago, about the time when Toynbee published the first volumes of his *Study of History*.[6] The main point, of course,

[5] Ibn Khaldūn, 1958 *The Muqaddimah: An Introduction to History*, translated from the Arabic by Franz Rosenthal, 3 vols. (New York: Princeton University Press, 1958).

[6] Arnold J. Toynbee, *A Study of History*, 12 vols (Oxford: Oxford University Press, 1934–1961).

there are self-controls in so-called 'primitive' societies. But they are of a noticeable different structure. In a so-called 'primitive' society, usually most of the time you are together with others. You are hardly ever alone. There is another balance between self-restraints and external restraints, which functions in a different way. That does not mean that there is no self-control, that there are no self-restraints, but they are dependent on the presence—more dependent on the presence of external constraints than the self-control structure of modern European people. So it is the changing balance between external and internalised self-restraints which is meant with the changing balance, which is meant by the directional diagnosis of Elias's book on the civilising process.

I especially remember an essay published in *Le Débat* on the Iroquois warfare[7] where there is mentioned a kind of escalation of—I do not know the English word—*Schrecklichkeit*, frightfulness or something like that, where the Iroquois people systematically trained their young men to develop a very high degree of self-restraint, self-restraint in relation towards physical violence, towards physical pain because if they are in a war with other tribes, they become prisoners, they are tortured, and they do the same with their prisoners and there is an escalation, a competition for not showing any sign of pain in the process of being tortured, so as to show the power and strength of the people of your own tribe. So you have certain areas—also mediaeval societies—you have certain areas where a very high level of *Askese*, asceticism, is practised. But you have other areas which, where the self-control and the degree of self-restraint is much different. You have feasts, you have opportunities where people get rid of everything that we consider to be self-control.

So the essentially directional character of the civilising process, understood in this way, is that the character, the structure of the self-controls changes towards a more middle line, to a stable, more even, more all-around self-control, steady and temperate and more or less the same in every kind of situation. It is more a tendency in a direction towards more autonomous self-controls in relation to what one could call more, relatively more heteronomous self-controls which are dependent on the presence of external enforcing agencies.

McNeill: I am very doubtful of the connection between civilised self-control and a simple society's self-control. It seems to me the varieties in simple societies are very great and the varieties of self-control within

[7] Joseph-François Lafitau, 'War among the Iroquois', *Le Débat* 5: 5 (1980), pp. 60–112.

different societies differ very greatly. I am very sceptical of the systematic superiority, if I may use that word, of the kind that you have accepted of the seventeenth-eighteenth-nineteenth century Europe.

Wallerstein: May I comment on this little debate because I am not going to take either side. I think it illustrates a problem. I think we know something about the nature of self-control in the modern world and I think with appropriate research we can discover a good deal about—not as much perhaps—but a good deal about the nature of self-control in what I call the various world-empires—in China, in Rome, etc. I am very sceptical that we know anything or even can know very much about the modes of self-control in what you are calling simple societies and I would call mini-systems. I think not only is our ignorance almost total, but I think it is almost structural. I find it very difficult to envisage how one might acquire adequate data to make reasoned statements about that. And I feel that there is a strongly speculative element in everybody's statements about this in which people infer from on the basis of what they think of as parallels to these simple societies in the modern world, which they may not be at all, parallels, and also a kind of a hidden social psychology which each of us carries around in which we say people would probably act in the following way in the following kinds of situations. It turns out that your hidden social psychology is different from his hidden social psychology and therefore you infer almost opposite. But I would caution much greater restraint about statements.

Elias: But what reason would you give for your scepticism?

Wallerstein: Well, because we have neither written evidence nor observational evidence of the actual …

Elias: Yes, we have.

Wallerstein: No.

Elias: We have. I have shown it. I mean, it is simply …

Wallerstein: What evidence do we have of how people in 8000 BC, in an area of Spain or of the Middle East or anything else, operated in terms of self-restraint versus restraint by other people? …

Elias: In 8000 BC neither you nor I have any records.

Wallerstein: That is right. OK. So what kind of evidence do we have at later points in time for people being described as not people operating within the Roman Empire and not people operating within the modern world, but people outside of that?

Elias: But why not take those?

Wallerstein: No, no, but that was the debate.

Elias: Why not take those of which you have evidence?

Wallerstein: Fine, but those that we have evidence do not bear the characteristic of being the simple societies. That is all I am saying. I think the statements about simple societies, what I call mini-system, are statements without empirical basis. They are inferential statements and I am objecting to them being made at all. I do not think we could ever demonstrate that either McNeill or Bogner was right about this.

Elias: You deny the anthropologists their bread.

Wallerstein: Yes, I would. I think it is important to do. That is to say first of all if you shall notice in the history of knowledge, the anthropologists, having recognised that their bread was based on such a feeble foundation, have shifted their area of concern in the last 20–30 years to that which is truly researchable. That is, anthropologists have moved into everything else in the social sciences—out of traditional anthropology just because of the weak basis. What anthropologists did historically—as a historical fact—is they went into areas of the world which had been brought into the framework of the world-system and they went with the guarantee of their personal safety as well as everything else—of external force. They observed how people behaved. There were two possibilities. Either they described what they saw then, which is how people in this zone operated, under the circumstance of their being involved in a larger system. Or they said 'Let us project back to a pre-contact stage' and they did that. By what kinds of evidence? By questioning various people at the present as to how they projected their image of, in fact, of the structure of this system at an earlier point in time. Now we have a good deal of evidence in fact that a lot of the early projections were manifestly wrong as demonstrated by later ones.

But in any case we also have a good deal of evidence from all sorts of other social sciences, that if you take reasonably intelligent people and you sit them down in a room and you ask them questions about what they think the situation was 30 or 50 years earlier, you are going to get projections in terms of the social structure of the moment, which are sometimes radically at odds with what an observer at that point in time could have observed. So, yes, I am denying the anthropologists their bread. I am making a very strong statement about the intrinsic limitation of these points in time that we call under the name of a simple society. And therefore, although I am very anxious to talk about long-term social processes, I want to observe the caution which you were preaching this morning which is that they can be empirically verifiable. And I do not think, in fact, that those statements are of a nature that could ever be—at least to the degree of the instruments presently known—it may be, it may be that as we progress with our skills in archaeology we will figure out ways of making more reasoned statements. And perhaps I would have to, 30 years from now, revise my view that one cannot make any statements. But at the moment I feel quite strongly that we are—it is wishful thinking to build a long-term analysis that involves our projections of that period in human time and that we are constrained therefore to make statements, which are verifiable about significant differences between behaviour in the Roman Empire and the modern world or between various other things. Those statements I find legitimate. But ones that go from early primitive *homo* or whatever—whatever term you want to use—and project a trend on, just lead us into this impasse in which you say one thing and McNeill says virtually the opposite and I sit here and I say, I have no way in which I think I could resolve that debate because I think they are speculative statements.

Evers: Just in case there may be somebody who has not grasped why Immanuel Wallerstein is so interested in destroying anthropology, let me point out that of course from his point of view the modern world-system has changed not only the facts of history but also the inner disposition so that if we from now try to look backwards, we must come to wrong conclusions because the former times were not a modern world-system in his sense. Now, of course, there may be other positions one might take, which might lead to the idea that anthropological evidence is after all valid, especially if it is gathered say around the seventeenth or sixteenth century and of course that is when fairly reliable—I hope fairly reliable—reports turn up. Not to speak of reports on tribal societies available in the, say the Chinese literature:

Wallerstein: That was written by the agents of the world-empire…

Evers: That is true. Well, at least we can agree that the difficulties are enormous.

Elwert: I am trained as an anthropologist but … I moved to sociology before this bad news came in. I am able to win my bread. I share scepticism on studies which present us present-day societies as the primitive, as stone-age economics. In fact they are present-day societies, articulated with the world-system as the societies on the fringe of the Chinese Empire were articulated with the Chinese Empire. So the reports of the agents of course do not report an original state of humankind. Nevertheless it is quite interesting: what are the differences they report and some of these differences point into a direction that the system of inner constraints existing at that historical stage is far more developed than one would guess from analysing an history of the core civilisations. This could be interpreted—I say could be interpreted—as a point in favour of McNeill's argument. But personally I prefer to go to situations where we have positive evidence and so I want to restrict my comments on three very limited points. Concerning the monopolisation of power—my evidence is only the formation of states in present-day Africa and this was a point you were referring to. I think we share evidence in this field. I was astonished about one omission. My evidence is that the process of monopolisation of power and of continuous construction of constraints for those exerting power is not only to be described as a process organised by those holding power. But it is also a process organised by subjects of power, by the subjugated populations. The fight against *Willkür*, against the arbitrariness of despotism, is as well a process coming from [lower] down. It is, so to say, a two-sided process.

A second point has to do with the *genius loci*. You may know that Arminius who fought the Romans successfully was probably fighting on this very spot in the Teutoburger Wald, and he was presented as one of the tribal people attacking the Roman Empire.[8] Well, you were quoting not Arminius, but you were quoting this reference to Roman history. I think it is slightly misleading because not only Arminius, but also the later attacks against the Roman Empire, came from so-called tribal groups which were formerly auxiliary troops which then transformed themselves into military tribes—but there is no clear distinction between military tribes and auxiliary troops. So what happened, in my view, in the decline of the Roman Empire was a loss of control over the military apparatus. Now I think this is a process which is very current today: the loss of control over the military apparatus, if we want to describe what is happening to the world-system now.

A last point: I like very much that innovation which you introduced to your system, to your theory by giving such a strong importance to the continuity of knowledge tradition. I think this is a very important point and this point—I think—has to be elaborated and the history of

[8] Arminius, or Hermann in German (18/17 BC–AD 21) was a chieftain of the Germanic Cherusci tribe who had earlier served as a Roman officer. At the Battle of the Teutoburg Forest in AD 9, he commanded a Germanic alliance which annihilated three Roman legions under the command of general Publius Quinctilius Varus. This is seen as a turning point in Roman expansion in the region.—eds.

institutions has also to be reinterpreted from this point of view. The most central element, and there I was astonished for the only scanty hints to literacy, seems to be the development of a culture of writing. A culture of writing, the transmission of knowledge through writing has a lot of structural prerequisites, structural conditions, which are not just the invention of an alphabet. There have to be institutions which give a condition for the maintenance of the identity of the system of writing, which fulfil the condition of a linguistic homogeneity in order that the concepts are the same on every edge of the communications system. There should not be breaks in the chain of tradition because no text can be understood just by reading. There has always to be a para-scriptural institution which introduces reading. So the caste of priests and scribes—and that is the caste of the people sitting here, as professors we are part of this tradition—is an essential part of culture and the maintenance or non-maintenance of this element of social organisation I think was crucial for the success or failure of systems as well as the other factors earlier described by you in your work.

Burguière: I want to only briefly to come back to this question about self-constraint and about Professor McNeill and Professor Wallerstein's scepticism about that. Because I think it is rather the key point of Elias's theory, what I take for the most interesting part of his theory. I think, that is, if we agree with this scepticism we will become a sceptic before the historical analysis in itself—if we say that we have no evidence to know what has passed before, [what people] used to think, used to do. We will come back to the concept of human nature or something like that. I think, for example, that the concept of self-control, of self-constraint has the difficulty of being confused with something like self-discipline, something like that. It is clear that in primitive, let me say 'holist' society (with Mauss's concept)[9] the self-discipline was very strong—perhaps stronger than in more complex and modern societies. But what seems to me to be interesting in the sociogenetic dimension of Elias's conception is the fact that the mechanism of internalisation with the new configuration of power not only gave to the individual the individual possibility to have better control of himself, but is predicated in itself to create a new conception of the individual—for instance a new sense of expectation, of strategy, the sense of dissimulation—and all these tendencies have been creative, not

[9]See Marcel Mauss, 'Address to the Société de Psychologie' (1923) [in *Oeuvres*, 3 vols (Paris: Éditions de Minuit, 1968–1974)]; *Sociology and Psychology: Essays*, trans. B. Brewster (London: Routledge & Kegan Paul, 1979).

only to promote new attitudes, but also to promote new cultural orientations and solutions. And this seems to me to be the real meaning of the dynamic dimension of self-control. And in this concept of Elias's I have envisioned to see a possibility of escape. I mean of progress for your concept of *histoire des mentalités*. Because it gives the concept, the articulation between mentalities and the configuration of society. But what I mean would be wrong, would be to use Elias's concept to try to make a sort of comparative history, for instance, using the concept of monopolisation, between empires of the ancient Middle East and contemporary empires. What they have in common is only the mechanism of monopolisation of power. But I would say that the effects of this monopolisation are so different, because the conditions are so different. … I think that Elias's concept is efficient to analyse a situation, a period, but not to keep the whole explanation of historical evolution. And from my point of view it is an advantage.

Elias: I think I should say to Professor Koenigsberger that figures like these here [pointing to the blackboard] have to be explained. It has to be explained that there are processes which transcend the epochs. And if historians are not willing to see that synthesis on an higher level will come, they must stay where they are. But it is without doubt that one way or the other—whether it is the way Professor McNeill goes or the way Professor Wallerstein or I go—there are too many facts which transcend the short-term perspective of historians. I mean simply to say it is so in the Middle Ages, but then it is totally different, means to disregard, to blind one's eyes to those facts which are not different. For instance what Professor Elwert said is very dear to my heart—the continuity and transmission of knowledge. There's a direct continuity of transmission of knowledge from old Sumer to our day. It's not only that words like *ager*[10] come from Sumerian, but many institutional transmissions go through the period frontiers. … If the training of historians makes them blind to these things, one cannot help them. Professor McNeill, I want to say I'm very grateful. I have not expressed the theme of constraint in agreement with my own feeling, my own knowledge. My own knowledge is—and that is what I will say very clearly in another essay which I'm preparing—'From signal to symbol'[11]—that self-control is a general human thing. But as Mr Bogner

[10] Latin for field, farm, land, and so on—eds.

[11] This appears to be a reference to the essays that eventually appeared just before Elias's death in the journal *Theory, Culture and Society* and posthumously as the book *The Symbol Theory* (Dublin: UCD Press, 2011 [Collected Works, vol. 13]).—eds.

tried to say, it is not difficult to see that the pattern of self-control, the pattern of self-control and the pattern of social control are different in simpler societies from those in more complex societies. And I'm really aiming at getting these differences clear. What emerges from there? It is, I think, an advance in social science knowledge to get together the structure of social control—whether it is that of a state or if you want of a market—and the structure of personality control, to bring them into contact with each other, and to see theoretically that they are dependent on each other. But I am grateful that you took up this point. I'm equally grateful for the fact that you took up the point about the nomads. I think I would agree that the difference is not very marked. That is to say, the differentiation between people living in early village states and people who are moving—the difference is not very great. I would put a question mark—and … this is a limit of my knowledge—[over] whether there are not in the fourth millennium already occasional migrating movements of simpler people. I am not absolutely sure of what you mean by the start of nomadry, which you date to the sixth century BC.

McNeill: Ninth century, I believe, ninth century.

Elias: I think that refers mainly to a specific development in the Russian steppe.

McNeill: It means riding horseback.

Elias: Yes.

McNeill: And all the adjustments that go with that.

Elias: It does not refer to the much earlier nomadic movements …

McNeill: No. No, not to Abraham.

Elias: … which you find from the Arabic peninsula or the Hittites and others. But in any case I stand corrected. You are right. I am not quite—I did not have your last point quite clear. So I will not, perhaps …

McNeill: Not important, not important. So just drop it.

Elias: I think we can discuss that among ourselves. I was grateful for the mention of Ibn Khaldun. I will try to see how that theory stands in relation to the present theory. Professor Elwert, I'm glad that this problem of transmission of knowledge appeals to me—I think it is a very important point and I'm inclined to elaborate. Theoretically, it has enormous consequences whether you stand in line of a tradition of knowledge of long standing or whether you do not stand in a society which does not stand in a [long] line of transmission of knowledge. I've forgotten the other point, I'm afraid.

Elwert: One was the control over the military apparatus.

Elias: I agree with you fully that controls are reciprocal. That was the point which you made. I am fully of the opinion that—and I think to some extent I have said it—that the pressure of the people on Louis XIV was a very active point which determined his behaviour. One of the central points of *The Court Society*[12] was really to investigate the power structure of an absolute monarchy and it resulted in the insight that it is anything but absolute—the power of the king. That he is in a great variety of ways dependent on the counter-control of the aristocracy, of the bourgeoisie, and of the people as a whole. So, that is the general statement I would agree, I would accept almost everywhere, it goes even for the slaves of Rome.

Professor Wallerstein, I may ask a question. Are you—if I understand your question rightly—are you suggesting a new kind of nominalism, according to which we, living in the modern world system, are constitutionally prevented from understanding people who are not living in the system?

Wallerstein: No. No. I am not. That is not what I have said. What I have said is that we can understand people who lived in ancient Rome or in China 5000 BC or in Sumer because we have enough evidence out of which we can construct reasonable perceptions of the operation of those social structures. What we cannot understand are people living in what get called under different names—mini-systems, primitive societies, etc.—because in fact we have no direct evidence or …

Elias: We go there. We do fieldwork.

Wallerstein: But they aren't any longer autonomous entities. The minute we go there and do fieldwork it is because they are part and parcel of a larger system, constrained by the operations of the system, and what we are seeing when we visit it is in fact how such a zone within the system—a very peripheral zone within the larger system—in fact operates. So it is a good description of a part of contemporary reality—that is an important part which we have to blend in. The anthropologists have a role. Their role is to help us construct the overall picture of the modern world-system. But their role is not legitimately to construct a hypothetical picture, which is what I think they were classically trying to do, of a structure outside that system. And it is just because of the lack of present-day … the only empirical knowledge we have, and that is very limited, is a certain kind of archaeological knowledge, and even that is very limited because it tends to be

[12] Elias, *Court Society*.

derived only from larger systems and not from the truly small system unit. You have up on there Peking Man with 50 people. Now first of all, I don't know how we know it's 50 people. It seemed perfectly plausible to me that it was 25–50 people. It fits in with my perception—inferential perception—but do we have even a shred of evidence—the most meagre—that it was 50 people and not 500? I suggest not.

Goudsblom: Well you could not have put 500 people in there physically.

Wallerstein: No, no, no. But you see perhaps it was 3000. I ask you another question. Do we know, for example, about Peking Man or any other such human, whether in fact, that social system of 50 people or 75 people or 200 people, lasted as a kind of sociological entity for 3 generations, for 6 generations, for 12, for 15, for 30 generations? Do we have any way of perceiving that? And I suggest, again, 'no'. So that if you then create a theoretical assumption about a directional trend that includes these hypothetical cases, I think it's on weak ground. If you want to give me a directional trend that says something about the trend of world-systems over time, from Babylon to Rome etc., then I say all right. You have some evidence and we'll discuss your evidence and maybe you're right or maybe you're wrong. And if you want to throw in, in addition, these so-called primitive peoples, I feel helpless empirically. And I think it's a mistake. It leads us astray.

Elias: Well, I can say to that: You are probably less sharp to anthropologists or to those who make hypotheses about simpler people than you are in relation to your own knowledge. I mean, you also make hypotheses where your knowledge …

Wallerstein: is limited.

Elias: … is limited. And so I think there is a certain right if we take the right to do things which seem plausible. It seems somehow plausible to assume that human beings, like the higher apes, first were assembled in very small survival groups and …

Wallerstein: May I just interrupt you. One thing. I now must bring in my own actual training. I went to Columbia University and one of my many teachers was Paul Lazarsfeld. And Paul Lazarsfeld—because your word 'plausible' rings an immediate bell—once wrote a marvellous article, one of the best he ever wrote, in which he started out by giving six obvious truths.[13] And any reader, including you, certainly me, nodded his head at

[13] Paul Lazarsfeld, '*The American Soldier*—An expository review', *Public Opinion Quarterly*, 13: 3 (1949), pp. 377–404.

each of the six obvious truths. He then said, public opinion research in the last 20 years has shown that each of these six obvious truths is in fact incorrect. And the 'plausible' worries me.

Elias: Well, I can only say that is a purely formalistic answer. Purely formalistic because it is difficult to assume that human beings could feed [themselves] in larger groups at an early stage. And it is also difficult to assume—it is quite plausible to assume—that they had to learn a technique of being gathered together beyond the great family level.

Wallerstein: I agree.

Elias: It is not unreasonable to assume that they were 50 before they could be 100 or 200. There is a certain sequential order, a certain developmental order in that. And I don't think, although I would not presume to say one can ascertain, Paul Lazarsfeld's ... highly formalistic assumption before hypothesis doesn't come into play here. The other thing I want to say is—I want to put a straight counterview. My straight counterview is that we are perfectly capable of going to simpler people, simpler societies—or, I should say, have been capable. It is now becoming very difficult. It is not the anthropologist which disturbs the picture. It is the very fact that simpler people today are gradually integrated into state societies. That is the factor which gradually completely disturbs the picture.

Wallerstein: It had already occurred when the anthropologists got there.

Elias: If Leakey makes a plea about his observation of bushmen and equates his newer bushman with the primitive hunter which lived 300,000 years ago, that of course is a thing which I would disagree with most strongly because a bushman today is already in a pacified area.[14]

Wallerstein: That's what I'm saying. Then we're in agreement.

Elias: This is a different thing from saying that anthropologists, by going there, disturb the picture.

Wallerstein: But I didn't say that they disturb the picture. I said that the picture they observe is an already disturbed picture.

Elias: If you mean that there are records about simpler society which are from the seventeenth century and even earlier and we can use these records very well to reconstruct a reasonable picture. I have no doubt that

[14] Louis S. B. Leakey, *The Stone Age Races of Kenya* (Oxford: Oxford University Press, 1935). The controversy about Leakey's views is clearly explained by Alan G. Morris in 'The myth of the East African "Bushmen"', *South African Archaeological Bulletin*, 58: 178 (2003), pp. 85–90.

we can do it with great precision—to reconstruct, say, the picture from hordes which lived 250 [years ago] to those today. I think one should not go too far. We have, in part, unused pictures and we know when they came and their bias because they were mostly by missionaries. But nevertheless, given their bias, they give us a very, very good and vivid picture. We can use that.

Evers: Well, I think we have to conclude the discussion without any final judgement. I must say, I was a bit disturbed by the line you took, Immanuel, in this discussion, because it sounded—and I'm sure this is a misinterpretation—it sounded a little bit like a plea for sticking where the evidence is and the evidence of course is with courts and kings, but the common people or the tribal societies we know only very little, so better forget about them. I'm sure this is a misinterpretation.

Wallerstein: It is. There's lots of evidence about the common people.

Evers: You said it—there is lots of evidence, but I would also say there is lots of evidence on tribal people. I remind you of empires that of course have—well, I think I'll stop. This is the best policy.

The Rise of the West as a Long-Term Process

William H. McNeill

Twenty-one years ago I ended my principal work with the words:

> 'The Rise of the West' may serve as a shorthand description of the upshot of the history of the human community to date. The rise of the west as intended by the title and meaning of this book is only accelerated when one or another Asian or African people throws off European administration by making western techniques, attitudes, and ideas sufficiently their own to permit them to do so.[1]

This seems as obvious today as it did then. Emergence of a new style of civilisation and of a new geographical centre capable of influencing the rest of the earth will take time, if, indeed, it occurs at all; and human lives are far too short for us to count on seeing any such shift come to pass.

[1] William H. McNeill, *The Rise of the West: A History of the Human Community* (Chicago: University of Chicago Press, 1963), p. 807.

W. H. McNeill (✉)
Chicago, IL, USA

A. Bogner, S. Mennell (eds.), *Civilisations, Civilising Processes and Modernity – A Debate*, Palgrave Studies on Norbert Elias,
https://doi.org/10.1007/978-3-030-80379-7_8

Nevertheless, from the vantage point of an individual lifetime and of the concepts available in that lifetime, as historians we must try to understand what happened in order to reduce the multiplex simultaneity of events as experienced by millions of diverse human beings to the slender proportions of grammatical discourse. Only by knowing what to leave out, what not to pay attention to, can such a feat be achieved. And only by accepting and then acting on a theory of social process can historians expect to have a criterion of relevance to guide them amidst the confusing plethora of data potentially available to their researches.

Selective transcription of what happened to be recorded in times past is not enough, even though many historians seem still to believe that ready-made truths about the past are to be found in primary sources, and need only to be transferred from the obscurity of the archives to the obscurity of learned discourse by a simple act of copying. But a collection of all the primary sources in the world would not record the whole truth. Experience is too multifaceted, too infinite for that. Whatever gets written down is itself an interpretation and drastic simplification of what actually went on. Moreover, what really happened can only be partly known at the time, since men are always ignorant of the full significance of their acts. Only subsequently—sometimes many centuries subsequently—can some meanings of particular acts such as the Crucifixion or Charlemagne's coronation at Rome become apparent.

There is also a statistical side to behaviour that remains aloof from, and largely independent of, individual intention and consciousness. Economists have long been aware that individual decisions to buy and sell add up to boom-and-bust rhythms of economic exchange—rhythms which no one plans but which nevertheless assert themselves even when no one desires it. Population dynamics exhibit other, less precise rhythms of unintended growth and decline. Cultural creativity, too, rises and falls in ways no one understands. These and other patterns of behaviour can often be recognised in retrospect, whereas contemporaneous observation of even the most familiar statistical patterns remains very difficult. Sophisticated economic indicators, for example, although devised by statisticians acutely aware of boom-and-bust patterns, are still not fully capable of registering the ups and downs of the marketplace as experienced in 1984.

A further complication arises from the fact that when a population learns to expect some sort of statistical fluctuation that seriously affects individual lives, then behaviour alters in anticipation of that fluctuation. Sometimes the effect is to exaggerate the phenomenon, as in the case of old-fashioned runs on banks; sometimes the effect is to damp back the expected fluctuation, as in the case of deposit insurance.

In either case, consciousness intervenes, directing conduct in the light of human concepts about the world and what is likely to happen in it. So changing ideas enter into and affect social behaviour, social behaviour affects ideas, and all the while innumerable human beings are acting and thinking and feeling variably from moment to moment, and recording almost nothing of their ever-changing state of being and consciousness. Yet surely it is the flickering flame of human consciousness that constitutes experiential reality—a reality buried and forgotten by each one of us moment by moment throughout our lives. Only fragments are ever written down or recorded in some other fashion; and such records are themselves selective, coded and interpreted representations of the reality, far removed from what actually happened, even in a single human mind.

If one takes a rigorous epistemological stance, therefore, writing a history of what really happened quickly becomes impossible. All we have is words and, perhaps, recorded visual images. Words are themselves embedded in languages; and languages are social products, permitting groups of human beings to communicate and thus concert their activities in common. The main function of words is to generalise experience, imposing categories and classes upon the flow of sensory inputs, and thereby allowing us to recognise useful objects, for example 'table', in innumerable and sharply different encounters with things. Beyond that, our words create the social world we live in to a very large degree, permitting us to recognise and respond appropriately to a policeman, a professor, a foreigner, or a fellow citizen as the case may be. Agreed-upon categories and classes of encounters with other human beings give meaning and direction to individual lives, tie us into human society, and constrain us to relatively predictable acts. Predictable individual acts become customs and mores in the aggregate. They harden into law in complicated societies when reduced to writing and enforced by kings and magistrates.

The survival value of such artifice is obvious. Predictable social behaviour is safer and more comfortable for all concerned and far more effective, inasmuch as under a regime of law and custom coordinated effort on the part of large numbers of individuals makes it possible to accomplish tasks otherwise beyond human capacity. Symbolic discourse thus gives human beings their extraordinary capacity to transform the natural world by collective effort. It is, indeed, what makes us human.

Historians are principally concerned with this semiological level of human interaction, and since words are what mainly sustain such behaviour, it is fitting and proper that historians should also rely upon words, interpreted from sources and strung together into discourse, to find meaning in the past.

But just as words, and concepts embodied in words, create the social universe within which humans live and move and have their being, so also historians must choose a set of words and concepts with which to give meaning to the past. To begin with, the obvious resort was to employ concepts at work in whatever society the historian happened to be a part of. This meant projecting local ethnocentric views of human conduct on aliens and ancestors among whom very different norms sometimes prevailed. History under such circumstances was Procrustean. Discrepant behaviour was either lopped from the historical record, or else exaggerated to show that those who behaved in such outrageous fashion were not really human. By writing local and chronologically restricted histories, puzzling incommensurabilities between expectation and recorded events could be minimised, of course; and by recording the triumphs of local ideals in action, historians could do much to stabilise group behaviour and define norms for subsequent generations. Such local ethnocentric histories could thereby carry the stabilisation of behaviour one step beyond what law could do, making conformity voluntary and enthusiastic and, perchance, more efficient and effective for being freely entered upon.

The role of defining and strengthening in-groups by codifying a flattering version of each group's particular past is, indeed, the principal social function that historians played, and continue to play in the contemporary world as much as in antiquity. For this purpose a historian needs only to master the language of the group whose existence and shared experiences he or she chooses to celebrate, though good historians and those who attained greatness regularly went beyond any given in-group's universe of discourse to consider the environing society as well. Thereby they tended to modulate the naive egoism and self-righteousness that all in-groups display ('us' vs. 'them') by recognising the shared humanity that runs beyond in-group boundaries and connects each group with other human beings. Such ecumenical histories are needed to prepare any given group for the buffetings of the larger world, in which others do not share its peculiarities and commonly distrust or dislike what loyal members of the in-group most delight in.

The familiar historiographic record shows this process in action clearly enough. In the nineteenth century, historians constructed glorious national histories for the various states of Europe and the Europeanised parts of the earth. Since World War II, a parallel enterprise is in train in newly independent states, but in the heartlands of the western world the most vigorous historical writing focuses upon experience at the private level—familial and demographic changes—or on variously oppressed subnational groups: peasants, workers, women, the insane, and so on. A smaller company of historians went the other way, inquiring into transnational themes and even aiming at world history in its entirety.

These new currents of historiography both reflect and reinforce the emergence of new group consciousnesses that in some degree challenge the national state. In writing the new histories, research scholars thus contribute to the continuing evolution of their societies, changing human consciousness in some slight degree by affecting the flow of messages that sustains and creates the human groupings within which each of us lives.

The shift towards subnational and transnational historiography remains far too weak to challenge the centrality of national-scale history writing. Yet within each of the leading national professional communities, the broadening of range inherent in these departures from the national frame puts considerable strain on the commonality of the historiographical enterprise. Historians who stick close to the primary sources and adorn their pages with ample quotes to demonstrate the accuracy of what they have discovered, do not seem to be doing the same thing that historians of civilisations and millennia are doing when they write world history.

The principal reason is that micro-historians can often move comfortably within a single, more or less coherent universe of discourse simply by reproducing words and concepts familiar among those whose history they write. Professional macro-historians, seeking to deal with human beings who espoused very divergent ideas about the world and how to behave in it, can scarcely do the same. Yet they cannot entirely abandon inherited language and concepts either. What they need is a specialised terminology, safely embedded in a flexible mother tongue, which will somehow be capable of ordering the experience of human beings in general, despite all the cultural diversity that human minds have elaborated across millennia.

Such a vocabulary has been slow to arise, and no one, I think, will claim that it has yet evolved a satisfactory scope and precision. Yet a real beginning has been made, and in making it macro-historians tend to pull apart from the company of micro-historians who study in-groups close to themselves in time and sympathy, whose local language therefore seems appropriate and adequate for recording the sufferings and triumphs that impinged on the consciousness of at least some individuals within the group, and which happened then to be written down. Such history, directly transcribed from primary sources though it be, remains inherently partial, and commonly abdicates any serious effort to locate the group experience within those larger social processes that are so dear to the heart of the macro-historian, but which remain absent from the consciousness of human beings as they actually experience life.

Here lurks the principal professional discrepancy of our age. The central notion for all varieties of macro-history is that of a social process (or processes) acting largely in independence of human awareness and so, by definition, not to be found recorded and awaiting discovery in some primary source. Yet it is worth pointing out that scale is not the decisive factor separating the two schools. Rather it is the difference between those willing to remain safely encapsulated within a given group's universe of discourse and those seeking somehow to transcend cultural boundaries. Ancient and mediaeval historians as well as historians of exotic lands regularly resort to terms and explanations that are far removed from the *ipsissima verba* of their sources, no matter how minute the subject of their inquiry. That, indeed, is what source criticism, the pride of our nineteenth-century predecessors, meant.

But when world historians come along and blithely use terms that distance them from the common assumptions of their own time and place as well as from those that governed the lives of people in the past, the validity of their procedures becomes questionable in a new way. Critical historians commonly use modern language to bring the objects of their study into conformity with our local, contemporary expectation and views of human probability. World historians are trying to perform a feat of intellectual prestidigitation, subordinating their own local social universe along with everyone else's to patterns and processes of which those concerned remain largely or entirely unaware. Yet insofar as they succeed and are attended to, the society around them will, in the changeable, intelligent human fashion, begin to alter its behaviour in the light of their new awareness.

How can historians or anyone else pretend to hold themselves apart from the social, intellectual, and cultural context in which they find themselves? How indeed? Frankly, I do not think it is possible, although one assuredly can stretch and adjust inherited concepts and expectations in ways that transcend at least some local biases. In particular, one may make room for human cultural diversity by devaluing the accuracy and adequacy of ordinary human consciousness, and supposing instead that human experience fits into some sort of newly invoked, precariously discerned 'social process'. Such tricks, played on the distant and the dead, are quite safe; but when the validity of a living group's self-consciousness and professed goals are called into question, the historian is likely to appear both arrogant and inaccurate to all who hold to local patterns of belief and to conduct based on that belief.

Why then does 'social process' hold such fascination for macro-historians? The concept, it seems to me, claims our intellectual allegiance because it recognises and allows for what men have always known: that what happens never completely conforms to advance expectations, while deliberately undertaken actions often provoke side effects that appal and dismay those who have to cope with the situation. In all traditional societies, wise men explained such discrepancies by invoking outside wills, usually the will of God or gods, or some lesser spirit. Sometimes miscarriage of an enterprise was blamed on the malevolence of mere human beings—the enemy, or some witch or traitor to the cause. But serious historians from Herodotus to Ranke and Toynbee[2] explained the course of history, with all its surprises, by recognising divine will acting among men in a fashion accessible only to faith.

I should stress that this view of the human condition is logically proof against criticism if one grants the premise of God's existence and His power to will and thereby govern the universe. Yet ever since antiquity, and more especially since the eighteenth century, many reflective persons have found the divine inscrutability an unsatisfactory resting place for historical explanation. They therefore developed the notion of social process as a surrogate. So far, I must say, the inscrutability of social process is not notably less than the inscrutability of Providence that it supplants. Yet it is to the anatomy of social process and its better understanding that I have devoted my professional career, and not without some sense of satisfaction and accomplishment. Let me try to sketch my view of what it is that I and other macro-historians are trying to do, and how we got where we are in anatomising this surrogate deity of ours—social process.

[2] Herodotus, *The Histories* (London: Penguin, 2014); Leopold von Ranke, *The Theory and Practice of History* (London: Routledge, 2010); Arnold J. Toynbee, *A Study of History*, 12 vols (Oxford: Oxford University Press, 1934–1961).—eds.

The starting point, I suppose, was the vision of a world machine generated by Galileo, Descartes, and Newton. If matter and energy behaved according to simple rules, might not human behaviour be similarly lawful? Adam Smith's unseen hand, utilising individual selfishness to enrich society, offers an early example of this line of thought, but it was only in the lee of the French Revolution, and in the minds of Hegel and Marx, that the concept of social process became fully generalised. Yet the realisation of freedom in the Prussian state that Hegel discerned in the record of history, and Karl Marx's dialectic of class struggle, were both firmly rooted in European historical patterns and left no real room for the non-European majority of humankind. Marx's slave–serf–wage worker sequence simply will not fit Asian, African, and Amerindian societies any more than Hegel's Prussian freedom did. This, it seems to me, was the crippling deficiency of these early efforts at anatomising social process in history.

Escape from hampering ethnocentrism came from two distinct directions. One was anthropological. With the establishment of European dominion over most of Africa and Oceania, a great variety of peoples came to the attention of European scholars, including many who seemed primitive to the anthropologists who studied them. Initial efforts to discover some simple evolutionary pattern among the earth's peoples went out of fashion—at least in the United States—in the 1920s. Instead, anthropologists undertook to study what they could see and handle directly: that is, material objects, customs, and folkways of Amerindian tribes and of some other isolated (mostly Oceanian) peoples.

Field anthropologists of the 1920s and 1930s were usually romantics. They hoped to find meaning and coherence among primitives in contrast to a growing incoherence in their own urban American moral universe, where the truth of inherited values was no longer obvious. What they sought they found. Anthropologists accordingly reported that the remote communities they had studied exhibited firm and coherent patterns of culture, varying indefinitely from one another, but each valid and liveable in its own terms.

The more intellectually ambitious sought larger patterns of coherence. Clark Wissler, for example, studied patterns of diffusion among Plains Indians by plotting the geographical distribution of particular items of material culture.[3] The resulting isobars could be convincingly interpreted

[3] Clark Wissler, *The American Indian* (New York: Oxford University Press, 1917); *North American Indians of the Plains* (New York: Smithsonian Institution, 1920).—eds.

as demonstrating a process of cultural diffusion northward from Mexico. Scientific as against merely speculative historical reconstruction of patterns of primitive life could thus be achieved, or so its proponents believed.

Another school of anthropology was less interested in historical reconstruction than in systematic description. For these scholars, time was trivial. What they wanted was an accurate typology of human society. Thus the spectrum extending from civilised to folk society that Robert Redfield constructed was in principle applicable anywhere, any time.[4] Since the characteristics of his two ideal types of society were abstract and general, they could be used to situate any particular society somewhere between the two theoretical extremes.

Contemporary American and European society, it must be admitted, figured near the extreme, 'civilised' end of Redfield's spectrum, being more secular, more diverse and incoherent, and more changeable than others. But such traits were costly. The enhanced power and wealth that civilised populations undoubtedly enjoyed had, it appeared, been won at a heavy psychological price—alienation, anomie, and personal isolation. The blithe assumption of older generations, to the effect that all humanity ought and would aspire to imitate the American and European example, no longer seemed plausible. Our civilisation, as exhibited by the western world of the 1930s, seemed an ambiguous good at best.

While this anthropological tradition flowered in the United States, a rather different line of development took place in Europe, rooted in the uneasy relationship between Frenchmen and Germans. In the late eighteenth century, a heartfelt wish to assert independence from French cultural models led Johann Gottfried von Herder[5] and other Germans to emphasise language as the natural vehicle for the expression of a people's soul. Different languages therefore created different cultures, each with a claim to unique value. Deference to older, more polished, but also more decrepit civilisations, such as that of France, was thus misguided. Germans (and Slavs!) should instead rely on their own folk spirit and shape their own culture in the hope and expectation of surpassing the French in due season, when natural rhythms of growth and senescence made themselves manifest.

[4] Robert Redfield (1897–1958); see his *The Little Community* and *Peasant Society and Culture* (Chicago: University of Chicago Press, 1960).—eds.

[5] Johann Gottfried von Herder (1744–1803), Enlightenment philosopher and critic.—eds.

This notion of plural cultures, each with a life of its own and predictable rhythms of growth and decay, was expanded to embrace the globe by Oswald Spengler.[6] Then Arnold J. Toynbee picked up the idea, partly from Spengler's pages and partly from his personal encounter with Balkan peoples and their alien ways as expressed in savage massacres between 1912 and 1922. Civilisation, which had been singular in French and English usage before World War I, thus became plural; and relations between separate civilisations became, at least potentially, subject to the sorts of analysis that American anthropologists had used to discern relationships among Amerindian tribes on the high plains.

This, in a nutshell, was my agenda in 1954 when I set out to write a world history. In effect I proposed to turn Spengler and Toynbee on their heads as Marx claimed to have done to Hegel; for oddly, it seemed to me in view of the record, both Spengler and Toynbee asserted that separate civilisations borrowed nothing of importance from one another, but instead lived out their separate lives in accord with some inner organismic life-force or state of mind. Material borrowings, however obvious, seemed trivial to these pioneers of global history, for they were interested in what Toynbee called 'the state of the soul' among those who shared in each of the great civilisations and disdained everyday material objects of the kind anthropologists had found so entrancing.

My approach, influenced by the anthropologists, assumed that borrowing was the normal human reaction to an encounter with strangers possessing superior skills. Even when borrowing proved impractical or undesirable, adjustment of existing ways of life might become necessary to protect against the seduction or threat inherent in contact with the strangers' apparent superiority. Such encounters thus appeared to me to be the principal motor of social change within civilised and simpler societies alike. A world history should, accordingly, focus special attention on modes of transport and the evidences of contact between different and divergent forms of society that such transport allowed.

Let me expatiate a little on my assumptions about the process of social change, before summarising the conclusions I came to and reflecting briefly on changes in my outlook that have occurred since 1963, when my book was published.

[6] Oswald Spengler, *The Decline of the West*, 2 vols (London: George Allen & Unwin, 1926–1928).—eds.

Social life, like physical change, requires some sort of force to set it in motion. Since contact with strangers provides the principal impetus to change, isolated human communities can be expected to settle down to a fixed routine by developing customs and techniques to meet all the standard situations of human life. In such a community, everyone would always know what to do without having to think about it. This, I believe, was the normal condition of hunters and gatherers throughout prehistoric times. Presumably inherited aptitudes in some profound sense fit us for such tightly knit, unthinking sociality.

Counteracting this normal stability is human inventiveness. This stems from play—play of the hand and eye and of the mind's eye, rearranging objects or words and symbols in some new fashion. Invention can be made for the fun of it and, occasionally, to see how some new scheme might work in practice. But inventions that catch on are more often associated with a time of crisis when some unusual circumstance imposes frustration on individuals or on an entire community. In extreme cases, when customary responses fail to achieve expected and desired results, random behaviour sets in—frantic and usually destructive. Yet once in a while a new workable pattern emerges from crisis situations—something that allows individuals or the community in question to exploit a new ecological niche, to relieve extreme situations by effective prophylactic action, or in some other way to improve chances of survival.

For a long time, probably, the principal crises human communities faced were ecological, and a long succession of inventions—tools, weapons, fire, clothing, and artificial shelter suited to all the variations of the earth's climate—resulted. But when almost all the habitable surface of the earth had been occupied, that is by about 8000 BC, contacts with other human communities began to become more important in provoking innovation. This was due to skill differentiation that set in as new ways of getting enough to eat became more and more important. Hunting and gathering—the initial mode of human existence—was gradually supplemented by rearrangement of natural landscapes through human action, so as to increase the number of animals and plants suited for human food. As this occurred, varying skills for producing larger food supplies arose. Some could be applied only in very specialised places, such as river flood plains, while others were applicable across much wider areas—slash-and-burn cultivation, for example, works well on almost any forested ground.

As varying skills diversified the life-styles of human groups, borrowing supplanted invention as the principal impetus to social change. After all, it is far easier to copy someone else than to work things out for oneself, and results are likely to seem far surer in advance. Contact between divergent human communities thereupon assumed the central role in provoking social change that it continued to occupy throughout succeeding generations down to our own time, when, all too obviously, the major impulse to continued technical change arises from the US–USSR arms race.

The process of borrowing and change went into higher gear after 3000 BC, when civilisations first started. These were societies favoured by an especially productive agriculture that began to sustain occupational specialists whose skills, through repeated exercise, quickly came to surpass those of ordinary, unspecialised food producers. Internal differentiation introduced serious friction to the body politic of civilised societies—Marx's class struggle. It also sustained higher skills, and the enhanced wealth and power such skills could create.

Superior wealth and power in its turn provoked envy and imitation among neighbours and neighbours' neighbours across hundreds of miles. The result was to set in motion complex patterns of invasion and conquest, of acculturation and of deliberate reinforcement of cultural differences in order to protect local practices from external threat. Overall, we see a whirl of cultural change, increasingly massive as time went by and as more and more distant human communities became involved in the process of civilised expansion—for that was the upshot of all that happened.

Civilisation expanded because most people most of the time preferred the enhanced wealth and power that civilised patterns of society conferred, and this despite the inequalities of status and income that specialisation involved, and the surrender or modification of old ways and egalitarian ideals that incorporation into civilised societies required. The result was that by 2000 BC the diffusion of civilised skills was altering human behaviour all across the Old World, from the valley of the Yellow River to Mesopotamia, and from Mesopotamia to India and the Mediterranean shores. Shortly before the Christian era, contacts across Asia became institutionalised in the form of caravans that traversed the entire continent on a regular basis. Interaction among the civilisations of Eurasia intensified accordingly. Then, after AD 1500, the hitherto independent (and less skilled) Amerindian system of interacting civilisations was incorporated abruptly into the Old World ecumenical system.

Nevertheless, local societies did not always crumble after contact with civilised polities or lose inherited patterns of culture just because they found themselves compelled or seduced into reacting to strangers with skills superior to their own. Selective borrowing was often possible. Such borrowings might fit into existing patterns of society with no very great difficulty, or might, on the other hand, precipitate a series of unanticipated further adjustments and readjustments. Severe and drastic periods of change brought about in this fashion tended to come in waves. Periods of unusually rapid change gave place to periods of comparative stability in an irregular and unpredictable rhythm. But civilised communities, being located at the primary foci of the whole accelerated cultural process, were never long at rest. Internal strains and external threats were perennial, and never allowed cessation of significant social change in each of the major civilised centres of the world.

Recurrent rhythms of state rivalry leading to unification of large territories under a single ruler are clearly perceptible within each of the civilisations of Eurasia and also of America. This was Toynbee's great discovery. But such empires were never stable for long; and eras of collapse and invasion not infrequently provoked fundamental new departures, thanks to fresh contacts, desperate invention, and eager borrowings by populations no longer sure that their inherited skills and attitudes were adequate.

This, in summary, was the social theory I used to make the writing of a world history possible. I simply set out to identify in any given age where the centre of highest skills was located. By describing them and then asking how neighbouring peoples reacted to such achievements, a comprehensible structure for successive periods of world history emerges, each distinguished from its predecessors and successors by a pattern of cultural flow outward from a dominant metropolitan centre. I need not rehearse details. Suffice it to say that I fixed upon successive efflorescences, beginning in Mesopotamia (3000–1800 BC), then in a cosmopolitan Middle East, taken as a whole (1800–500 BC). Next the Aegean basin and Mediterranean shorelands rose to primacy (500 BC–AD 200). Then it was India's turn (200–600), only to be succeeded by the Moslems, based in a reintegrated Middle East (600–1000). In the most recent millennium, world leadership oscillated towards the extremes: first the Far East (1000–1500), and most recently the Far West (1500–1950).

Each era of efflorescence, after the first, involved a preparatory period of large-scale borrowing from more accomplished cultures—usually from the civilised centre dominant in the immediately preceding era. But such borrowed elements entered a distinct institutional and cultural setting, and thereby characteristically attained new scope and importance—often of a far-going kind. (Consider Europe's use of Chinese gunpowder and printing!) This, along with a continuing flow of new invention, and drastic efforts to accommodate old and new in some supportable symbiosis, was what provoked each period of efflorescence and allowed some new region to forge ahead of other peoples in defining a new, powerful, and attractive way of life.

Such perspective makes Europe's recent global dominion simply the most recent example of a recurrent phenomenon. To be sure, intensified communications and cheapened transportation meant that the impact of European civilisation on other parts of the earth was much more severe than anything that had ever happened before. Local autonomy of the kind that used to prevail, as between each of the principal Eurasian civilisations, is now clearly diminished by the density of contemporary communication. Perhaps the processes that used to reward a clustering of high skills within a relatively small geographical compass will be modified by instantaneous electronic signals. But as long as cities form and governments govern, it seems likely that centres of cultural dominance and cultural slopes towards provincial backwardness will continue to exist among us. If so, the long-standing pattern of sporadic shift in the location of dominating metropolitan centres may be expected to continue to manifest itself among the great cities and civilisations that share the management of human affairs today, as in times past.

Since 1963, when my book was published, two modifications of my views have achieved a modicum of precision. First of all, I am more conscious of the slow emergence of our contemporary ecumenical society than I used to be, and less satisfied by the notion of separate civilisations as the key concept of world history. In particular, the autonomy I attributed to the major Eurasian civilisations until 1850 ought to be modified by recognising an emergent ecumenical cosmopolitanism, starting with the establishment of caravan connections before the time of Christ. Ecumenical cosmopolitanism attained intensified significance from AD 1000, when China and western Europe alike began to participate in a sophisticated world market far more intensively than before. It looks to me now as though the ancient history of Mesopotamia, Egypt, and the Indus civilisations, which subsequently merged into a cosmopolitan Middle Eastern civilisation, was in effect recapitulated on an enlarged geographic scale after 500 BC the principal actors having become China, India, and Europe, in addition to the Middle East.

This way of thinking about the history of the world emphasises communication even more than I did before, and reduces the separateness of the world's civilisations more drastically than I did when writing *The Rise of the West*. But communication is what makes us human, and civilisation is no more than a shorthand summation for myriads of messages exchanged among large populations and uniting them in fundamental beliefs and expectations. Changing traffic patterns define changing social groups, and there is nothing sacred about 'civilisation' as a unit. Historic civilisations were, perhaps, no more than a register of the dimensions of communications systems based on writing (sacred scriptures above all) in combination with animal transport and sailing ships. New modes of communication may therefore be expected to create new patterns of human association, and the old civilisational building blocks may, in time, erode away. Who knows?

The second considerable modification of my views is this: I now believe that human culture is itself set in an ecological context that evolves in partial independence of cultural developments. This is notably the case with infectious diseases, as I argued in *Plagues and Peoples.*[7] Climate is another significant variable which I had previously overlooked. Deforestation, pollution, and other large-scale changes in the ecosystem may yet prove to be a third. The ever-changing ratios between human numbers and resources, as currently available for the maintenance of human life, in accordance with a particular technological and social structure, constitute still a fourth at least quasi-ecological limit upon human affairs.

More generally, human life and society, it seems to me, can best be understood as resting within a hierarchy of equilibria, each of which is itself dynamic and susceptible to unending and occasionally to catastrophic change. First is the physiological—an equilibrium among molecules. Next is ecological—an equilibrium among organisms. Last is semiological—an equilibrium among symbols. Human cultural behaviour operates at the level of symbols, of course; and how we behave from day to day depends primarily on symbolic messages in and messages out that define our consciousness. The symbolic world is complicated enough in all conscience, but the interactions across these levels of equilibrium are far more difficult to understand, powerful and perpetual though they be.

Clearly, symbolically directed behaviour channels our intervention in the ecological equilibrium; and experience of what does happen when we interact with the natural world feeds back into our vocabularies and concepts in such a way as to alter and correct, refine and redefine the symbols

[7] William H. McNeill, *Plagues and Peoples* (Garden City, NY: Anchor, 1976).

we use to guide our further behaviour. Practical applications of natural science stand as evidence of how powerful this process can be; but in the deeper past, theological and magical ideas guided human actions just as efficaciously as science has done in recent times. Moreover, rules about when to plant or how to guard against disaster were corrected and adjusted in the light of results, just as we do today.

But, as I said before, symbols also define human roles and groups—friend and foe, relative, stranger, trading partner, fellow believer, fellow soldier, fellow citizen, and the like. The rise and fall of groups within a civilisation and among civilisations is the substance of human history. Much is self-contained: that is, created by the flow of messages in which our conscious lives are immersed. Yet there are constraints and pressures on our equilibrium of symbols that come from without. Some messages work, whereas others fail to achieve satisfactory results for those who act on them. These constraints arise from nature—animate and inanimate—and also from human nature: that is, from the ecological and physiological levels of our lives. No one, I think, is able to handle these hierarchical equilibria interactions very comfortably. Yet there are interesting convergences between these speculations about society and those which brain physiologists have advanced about how the cells of our brains produce consciousness and memory. Perhaps someday thinkers will be able to put each level together with its fellows in a far more satisfying way than I am able to. If so, a firmer theory of social process, and with it of world history, will become attainable.

Short of that, and we are indeed far from any such synthesis, it nevertheless seems obvious to me that macro-historians, in seeking to anatomise social process more accurately, are pressing forward on one of the more active fronts of intellectual invention in our time. This is the first or, at most, the second generation that could seriously attempt world history, if only because it is within this century that scholarly description and analysis of the various non-western societies began to achieve globality. It is, likewise, the first generation in which sufficient detachment from naive ethnocentrism has become possible, largely because the self-confidence of western peoples was so shaken by World Wars I and II, and by the fearful prospect of World War III.

To find oneself in such a position is both exhilarating and oppressive. We may be in process of formulating views that will seem classical to generations yet unborn. We may simply be brushed aside by dogmatic simplifiers in search of a simple orthodoxy, enforceable through police power. Or we may merely become quaint as our ideas and formulations become completely outmoded through the further evolution of human knowledge and belief. Time alone will tell.

Discussion of McNeill's Paper

Transcript

Wallerstein: May I ask a question about a statement? What is not clear to me from this presentation insofar as you began—which I thought was very helpful—with your own intellectual history, and you have the image of a diffusion of cultural traits across a series of successive—some contemporaneous and some successive—civilisations, and you've talked of the permeability of civilisations, so that I have a kind of image in my mind of some kind of wiggling phenomenon, moving on through these civilisations. Does that imply a kind of evolutionary presumption that as it moves through somehow there is a kind of cumulative …?

McNeill: Yes, yes there is. After I finished my book I sort of looked back and I said, 'My God, I've written the history of power—wealth and power'. And I didn't think I was doing that but in the end I decided I had been. And I would justify or apologise for this by saying that most people most of the time do prefer wealth and power against their alternatives—poverty and helplessness. And this kind of option has accumulated across time. Certainly there's been a technological cumulation. And not just technological. Also I should think institutional cumulations that make us more powerful and more wealthy than our predecessors were. And the

Transcript (✉)
Berlin, Germany
e-mail: Artur.Bogner@web.de

A. Bogner, S. Mennell (eds.), *Civilisations, Civilising Processes and Modernity – A Debate*, Palgrave Studies on Norbert Elias,
https://doi.org/10.1007/978-3-030-80379-7_9

general option of humanity, given such choices, as best they perceive them, is for wealth and power more often than not. So every so often people jump off a cliff and say 'No', but very rarely. So then that effect is an evolutionary one. I do belong in the nineteenth century. I appear to accept the description of a child of the Enlightenment. Even if you don't like it, I am a child of the Enlightenment and don't really apologise for it.

Bertelli: You're right that your approach … assumes that the borrowing was a normal human reaction to an encounter with a stranger possessing superior skills… May I ask you, about just another two examples. One is the Roman versus Greece and the other is [the Lombards] versus the Romans, too. May I ask if you understand the meaning of 'acculturation' as a one-way process?

McNeill: No.

Bertelli: Because this is what is implied here.

McNeill: In any cultural encounter there is reciprocity of sorts, but if a stranger comes in with skills that are recognised, that appear to be superior to that which I possess—what he or she can learn from me is probably not very much—what I can learn from him or her is far more important to me. So that in any given encounter the cultural flow, the change of behaviour which ensues from the cultural encounter may by quite lopsided. But there will still be some residue on their part. I think always. I think there's always a reciprocity. But possessors of higher skills are likely to be indifferent or condescending and say, 'I know everything that's important already'. So the European trader on the coast of China, for example in the nineteenth century, could live amongst the Chinese for 30 years and not even learn Chinese—the language. And certainly not—his behaviour is still modified by the fact that he had Chinese servants and goodness knows what. He wasn't the same man that he was when he arrived as a young fellow at 20. But his borrowing from the Chinese was minimum. Whereas the Chinese, faced with this all-powerful taipan, had to take a base of action, had to do something about it. And the whole history of nineteenth-century China revolves around 'What in hell do we do with these foreigners, about these foreigners?' So you can have very lopsided exchanges, but I think always reciprocity, I would presume … No, I appreciate that correction—or acute elaboration.

Elias: Professor McNeill, I just hope that in the discussion we could clarify a few basic concepts. Because we like precision, not only with regard to historical data, but also with regard to the conceptual apparatus which we use. So I am not clear, and I would be very grateful for elucidation, what you actually mean by 'civilisation'. I mean—in what relationship does the term 'civilisation' stand to the term 'society'? For me the subject matter of history—that which has history—are always human beings who have history, or groups of human beings. And in my particular language, which does not mean very much, usually 'self-ruling groups' are the subject of historical changes. But in what relationship—that's the question, one question—does the concept of 'civilisation' stand to the human beings who are the subject of this thing? And that goes together with the next question. What exactly do you mean by 'ecological'? I mean in some of your writings it almost seems that you use—but I do not want to put it in because you did not, I think, here very much but maybe it will be illuminating for me to see—you use the term 'parasite' for taxpayers as well as for bacilli.

McNeill: No, tax-collectors,

Elias: And so I would be grateful for an illumination for the—of the word 'ecological'. Does it really apply to the state as well as to parasites? And does 'civilisation' apply to human beings or is that something outside the human race?

McNeill: Oh, no. Civilisation is certainly an affair of human beings. The best definition of 'civilisation' that I have achieved for myself is a style of life. Analogous to artistic style, Something that you can recognise when you encounter it as *sui generis*. It has a certain loose coherence of its own. In practice I must say, it seems to me that the term that I have, the delineation of civilisations, very largely revolves around this notion of a sacred scripture, of a religious system which provides certain norms of behaviour and aspiration. Norms of aspiration I think I should say rather than norms of behaviour. So that the great historic civilisations of Eurasia were from, say, well from the fifth century onward, very largely defined in terms of certain religious groups. Now, not earlier than that ... A civilisation, I believe, is built upon certain biases, certain presumptions. It is a certain phenomenon, learned behaviour. And certain, as it were, master institutions, master ideas, around which the other aspects of the cultural behaviour are adapted, adjusted, arranged, produce the civilisations. A civilisation's articulation is relatively loose as compared to a folk society's, a much tighter cultural definition of behaviour. And subgroups within the civilisation may differ very strongly from one another. Often hostile relations, struggling relationships between different social groups, different social classes, different ethnic groups may be all folded into a civilisation. But what unites it is this shared aspiration.

Elias: May I just ask a supplementary: Do the norms and ideas which you say are very central, or religion—do they hang in the air, independently of the human beings?

McNeill: No, by no means. By no means whatever.

Elias: I beg your pardon.

McNeill: Not at all!

Elias: So how do they—are they in relation to human beings in groups?

McNeill: Well, I must say that this depends on which civilisation, which religion we're describing. But there will be institutional forms and particular bearers of the given high cultural mode: priests, clerks, persons who embody the written … There's a certain stabilisation in writing, also that comes with the early beginnings of civilisations. They're built around priests—as you were saying—first, priest civilisations. And it is that stabilisation in form of written texts, sacred texts, which gives a sort of cohesion and endurance, it seems to me, to civilisation, a style of life.

Elias: Have simpler people civilisations?

McNeill: I would rather not use 'civilisation' for simpler peoples.

Elias: Why not?

McNeill: This is a confusion of words which has become very popular in the United States—because it is thought to be undemocratic to deny civilisation to what I want to call simpler peoples. And I also like the word 'barbarian' because I think it means something real in terms of communities that are in contact with a civilisation and recognise the civilisation has skills and can do things they can't do. But they also treasure and hold precious those things which differentiate themselves from the civilisation. There is a very deep ambivalence in their relationship to the civilisation, built into the civilised–barbarian encounter. And it's unfortunate, 'barbarian' has the other meaning of 'uncivilised', of derogatory meaning …

Bertelli: Perhaps they were differently civilised, who didn't mind the superiority of one against the other?

McNeill: The term that was invented in English to be neutral was 'culture'. They all have cultures. Barbarians have culture. Civilised people have culture. Barbarians have culture. Civilised people have culture. And my isolates, my 'primitives', my—the word 'primitive' I want to avoid because I think the view that most society is primitive, they are all just as old as the rest of us. There you had the same chance to experience the world as the most civilised people. And it's not 'primitive'. It's 'simple'. Simple societies, barbarian societies, civilised societies all have culture. And that's the English usage. 'Culture' is the non-pejorative common trait. But that there are different kinds of cultures—the polarity that I inherit from Mr Redfield[1] between civilised and simple seems to me real and as a middle term I want to have 'barbarians', because I think barbarians occupy a distinct niche in this cultural slope pattern from metropolitan centre to the people who are sufficiently isolated that they are in no significant enduring relationship to the high cultural centres. The reality is a message system, a communication [net]. That's what holds civilisation together. That's why I think these shared aspirations and the written text, the sacred text are so important as defining the boundaries of a civilisation. And what makes it human is messages in, messages out. This world of symbolic signification and the coherence of common behaviour, of collective behaviour organised around those symbols. And this is culture. That is what cultural behaviour is. And every community exhibits this. It's what distinguishes us from the beasts. Now ... the question, what ecology means? Ecology, it seems to me, is the series of relationships with the natural world which ... occur independently of symbols. Disease organisms is one. The changes of climate, other kinds of changes of the natural environment that have powerful effects may exert serious constraints upon human culture. We're accustomed to doing it that way and presently that natural road is altered in some fashion that doesn't allow us to do it in the same old way. And this ecological level is quite pervasive and quite thorough. It sets limits. At the same time our culture allows us to intervene in that natural world and transform natural landscapes in different ways at different times so that we have ... this hierarchy of equilibria. And the interrelationship, the feedback relations between the different levels of equilibria are extremely complex and I do not feel that I have it well specified, well described in my own mind. The use of 'ecological' to—I mean my metaphorical use of the tax-collector as the equivalent to a predator in human society. I did say that, but I also said I'm not going to use that metaphor in the book I just wrote, *The Pursuit of Power*, because I think it is more distracting and more destructive to clarity of thinking than helpful.[2] And so I backed away from that. In *Plagues and Peoples* I did say that.[3] There's a little passage where I make the comparison which you may have read. But I think it is a confusion of language which is better not get

[1] Robert Redfield: see note 4, p. 107 above.—eds.

[2] William H. McNeill, *The Pursuit of Power: Technology, Armed Force and Society since AD 1000* (Chicago: University of Chicago Press, 1982).

[3] William H. McNeill, *Plagues and Peoples* (New York: Anchor, 1976).

indulged in. And I apologise for misusing the term 'predator' in that fashion, 'macroparasite' in that fashion. Because it is a metaphorical expansion of the term and it's really confusing rather than helpful. You don't think so?

Wallerstein: No, I thought it rather good.

McNeill: The difficulty is this, that the predator is predatory within the semiological system, within the cultural system. And that confuses two levels of organisation—the ecological with the …

Wallerstein: It has all the virtues and vices of any analogy

McNeill: Okay, okay. Well, that's why I thought it was better not to use it.

Wallerstein: It made your *Plagues and Peoples* seem much more relevant, frankly to everything else that everybody else is doing, by just using that metaphor, and therefore indicating that there were different ways of being parasitical.

McNeill: I'm glad you liked it. Perhaps I shouldn't have backed away from it, but I think it's capable of being radically misunderstood. That's the difficulty.

Goudsblom: Yes, I'm afraid it's a child whose legitimacy you can no longer deny. One of the best comments that have been written about your work, by the economic historian Eric Jones, also uses the term 'macroparasitism'.[4] …

McNeill: In one of my nutty moments, I do like it. The problem is, what I just said, it confuses two levels of organisation. And that's tricky, that's dangerous.

Goudsblom: No, I think apparently your scientific conscience got the better of your political conscience.

McNeill: Perhaps.

Brands: I have a very old-fashioned question referring to a very parochial view of which I am genuinely ashamed. Just asking about that period which is called 'the period of the double revolutions', *die Sattelzeit* [Saddle Age] I think in German, of the industrial and of political revolutions around 1750, 1850. My question is about your terminology. You said a lot in general. All the things, I mean communication, transport and so on. It has been all there for a very long, long time. And a couple of times you said the word: 'Of course the impact was different'. I mean far more

[4] E. L. Jones, 'Economic history at the species level', *History of European Ideas*, 3: 1 (1982), pp. 95–105, a review of McNeill's book *The Human Condition: An Ecological and Historical View* (Princeton, NJ: Princeton University Press, 1980).

overwhelming later on than in the beginning. But it was not a difference of kind. ...

McNeill: Well, this is, of course, a matter of taste and judgement. I think I do react against the emphasis upon the uniqueness of modernity. And my habit of mind as a historian is to say, 'Yes, but there are continuities—that are also significant or even more important!' And the notion that, with industrialisation, urbanisation and the development of mechanical transport and now instantaneous communication, somehow that everything that happened before 1750 is no longer real. It runs against some grain in me. I mean I like to think that humanity is humanity and the cultural processes and social change has not taken that sort of quantum departure from the past. Now, there are more people who think the world began in 1750, 1850—they usually think it began in 1950—than there are who think there was a world of people in the past which has continuity. Let me say in defence of the proposition that the world is ... that there is a continuity of the world. Most of the world is still rural. Is still peasant. Most of the world is still a peasant world. Our urban concentration is a kind of eccentric departure from the norm. We haven't changed that much. It's only if you project upon the rest of humanity in the next, I don't know, 50 years, 100 years the same urbanisation, the same development of industrial production, which I think intrinsically implausible, that the breaks with that past are going to be as catastrophic as they have been in the American Middle West. We have indeed broken with our peasant past in North America. You haven't done it completely in Europe. You're working hard at it. You haven't done it all the way. And the rest of the world—is hardly just beginning.

Wallerstein: Footnote: The statistics of just the last 20 years show a remarkable rise in urbanisation, but then you would have to make your break that of the last 20 years and not 1759.

McNeill: And the lands are still full of peasants. But they're moving into the cities ...they have a lot of countryside.

Wallerstein: There's a sudden absolute ... upsurge of urbanisation although one might question whether that's urbanisation by making many people moving to the cities.

McNeill: So, I think there are arguments that we are in this process of social change that has continuities that run before 1750.

Stauth: I have a very short question. Just in terms of where you noted Europe as still being a type of peasant agricultural society. What do you think Saudi Arabia is?

McNeill: It's not peasant.

Stauth: In terms of your notion of civilisation and specifically in terms of distinguishing the two wealths—wealth and skills.

McNeill: It is surely a part of the Muslim world. Now, whether you accept the notion of Muslim civilisation or say the differences between the different sects of Islam are so intense that they don't share—I think they do share enough to call them a Muslim civilisation. I've written that in my answer. Now it's been transformed and stretched and broken in Saudi Arabia in such a degree that explosion tomorrow is entirely possible.

Evers: You get into trouble when you get to India that way.

McNeill: Do I? I thought India was a civilisation also.

Evers: Is it a Muslim Civilisation?

McNeill: No.

Evers: A Hindu? Sikh, perhaps?

McNeill: Well, I would call it Hindu. Hindu in a very loose sense, in which there was first a Muslim overlay and partial withdrawal. And then a British overlay and much more drastic withdrawal. And certainly a kind of a mix …

Evers: This is a problem with historians. They always withdraw into facts when they are asked questions about concepts.

Elwert: I'm triggered by the concept of simple societies. Who is simple, who is complex? Judging from the ongoing expansion process, it rather seems to me that the success of capitalist society is linked to the reduction of complexities, to the structural simplicity. Money as a means of exchange seems to be rather simple and simplifying. The state as the centralised organisation of coercion seems to be extremely simple as compared to the means of social control in the so-called 'simple' societies. And last not least the use of writing, which is the most powerful reduction of communication to just words and symbols as compared to systems of communication with sound, smell, visual expression and so on, which we have in other cultures. So it seems to me that the strength of this type of society and civilisation and culture we are living in is linked to simplicity. So this is a remark just to terminology. Maybe we should not be too serious in all these associations we use and in fact [this term] 'simple' is a metaphor and it has all the drawbacks of metaphors.

McNeill: Well, in my usage I think there is a relatively clear meaning. I hope there is. A simple society is one in which there is little professional specialisation, occupational specialisation—which most people do the same thing.

Elwert: Well, I don't want to go into details.

Bogner: That's the problem with sociologists. If they are asked about facts, they go into conceptual sophistication.

Elwert: It is obvious that this joke had to come, and the point which one might raise is the point of the structural complexity of the society where within the individual the combination of different skills is very high. And the problem of organisation which is caused by the multiplicity of paths and roles, which are organised within the individuals—this is a very specific problem of social organisation and some anthropologists are very conscious of it. Well, but I don't want to expand now on this aspect of structural simplicity, but I think any discussion of commodification in fact brings us back to the strength of at least one process of simplification. There's another point which I like very much in your elaboration here which is very frank also in the development about your [own new] work. The deliberation, of course, about the skill of writing. Should this not bring you to a different station of different types of skills. I think there are civilised skills, as you call them, and civilising skills. Writing is a skill which allows for the diffusion of skills. And this I think has implication also for your general concept of diffusion. After the introduction of writing, diffusion became a much more powerful means of getting new ideas, of getting innovations, than creativity. This poses, of course, the problem for empirical research that it is very difficult to have knowledge about creative innovation outside the written cultures. Of course, you make statements about these. We have just one example about which we know because of archaeology. That is, the innovation in the nomadic societies outside the centre—outside the centre to which you give so much importance. The introduction of the saddle allows to ride on the horses, and on the camels, and not to be drawn by them which was much slower. The introduction of the tent which is a very recent innovation allowed for this very high mobility of the nomadic groups as compared to the use of the yoke. Now, we know just about these two innovations by some accidents of history. Some written reports, some findings of archaeologists. But it's very difficult to know what were the type of innovations in those cultures which did not have writing. So I think writing about written cultures one has to be diffusionist, as you are, because writing gives such a competitive advantage for diffusion of innovation as compared to creativity. But expanding diffusionism to the so-called 'simple cultures' about which we lack of knowledge might be an ethnocentrism.

McNeill: Have you read Clark Wissler? This is an American anthropologist who did it … about the American tribes of the Great Plains.[5]

Elwert: We had in German anthropology a lot of discussion about diffusionism and just to say this—there was a lot of blunder in it.

McNeill: Oh, I understand that.

Elwert: And I'm very sceptical …

McNeill: Without written documents you're in real trouble and this is what Clark Wissler thought he had overcome—largely by attestations of … remains and … reports and photographs and artists' drawings. It's an interesting example. I think it's quite a convincing case of diffusion amongst the Plains Indians, who were originally …

Wallerstein: Every knowledge we have about the Plains Indians which is positive knowledge that I can accept dates to a period where we can speak and have to speak about the world-system.

McNeill: Yes, yes.

Elwert: And the culture of the Plains Indians and all these others has to do with the introduction of trade using not necessarily money but using equivalents of money. This is one of the transformations we spoke about earlier. You have a transformation of the internal structure because of an outside transformation of the economic system on the fringes so this …

McNeill: It's my vision that this is what's been going on since civilisation first arose—interactions of this kind—that they all have been parts of world-systems. Whether there is a state, empire. Only part of the time in the pre-1500 world were there world-empires. Often there were a multiplicity of states, you know that, so that the difference is not—seems to me not—great from the modern world-system. Just as Europe's the centre now—or was—till 1950 anyway, and in earlier times it was somewhere else that the major centre existed. And, of course, within Europe you get the sort of shift from the Mediterranean to the North Atlantic, that M. Braudel has made such a lovely analysis of.[6] It isn't strictly a monolithic entity. In China, there's a movement from the Yellow River to the Yangtze River which is quite apparent. For some dynasties, the Yangtze River … south instead of in the north. So this kind of migration within a general civilised area is also, came also to occur. And my 'simple' societies that are reacting

[5] Clark Wissler, *The American Indian* (New York: Oxford University Press, 1917); *North American Indians of the Plains* (New York: Smithsonian Institution, 1920).—eds.

[6] Fernand Braudel, *The Mediterranean and the Mediterranean World in the Age of Philip II*, trans. Siân Reynolds, 2 vols (London: Collins, 1972).—eds.

to encounters with strangers seem to me to be doing what the Plains Indians did in the nineteenth century, since there were significant differentiations of skill amongst human beings which takes us back—as I say—to the beginnings of civilisation. The idea of an unspoiled primitive who never met a stranger seems to me to be a total myth. We can't know about them by definition, but I don't think they ever existed.

Koenigsberger: Since I seem to be the only microhistorian around, perhaps you'll forgive me if I come back to this problem once more. Now I accept everything you say. I think, I'm entirely with you. I think this is fine. What I'm wondering about is—is it anything more than descriptive? Where I'm bothered is that when it comes to claiming for this—and this is where I'm bothered with Norbert Elias, too—claiming for what seems to me to be descriptive, the status of either a law, *Gesetz*, *Monopolgesetz*, whatever or of a model. It seems to me that I accept entirely that societies or civilisations, whichever, are permeable, that they are constantly influenced from each other—but now, how much? How important is the influence in particular directions, at particular points, how important is it for the further development of the society? You, yourself, said—if I understood you correctly, but please correct me if I'm wrong—that you find it very difficult to find a balance between these different equilibria of the different conditions which you have produced. Speaking now as a microhistorian I have myself tried to do this in a very much more restricted field—that precisely where I first used Norbert Elias's categories in the field of the development of European parliaments.[7] My conclusion was that it is, although one can produce a descriptive analysis of the development of parliaments and can see all the various forces which impinge upon why some parliaments became important or rather established themselves against their monarchies and why others did not. And I was very much aware of the permeability of all the societies in which this happened. My conclusion, however, was that it isn't even theoretically possible to analyse sufficiently all the forces which go into determining what the outcome is going to be. Now, having found in microhistory the impossibility of doing more than an approximation, at best, of coming to... of producing a workable model and not just a description, I must say I have not been

[7] H. G. Koenigsberger, '*Dominium regale* or *dominium politicium et regale*? Monarchies and parliaments in early modern Europe', in P.R. Gleichmann, J. Goudsblom and H. Korte (eds), *Human Figurations: Essays for Norbert Elias* (Amsterdam: Stichting Amsterdams Sociologisch Tÿdschrift, 1977), pp. 293–318.—eds.

convinced by anything I've heard today—and it has been absolutely fascinating—that it is possible to do so for civilisations or societies on a worldwide scale and over the whole of human history. I still remain unconvinced.

McNeill: Well, I certainly don't think I can determine why this particular event occurs here instead of there and so on. The word 'determine' would be much too ambitious. And in that sense it is description. At the same time the question of how important a borrowing process can be seems to me relatively easy to illustrate. In the case of modern Europe, think of what the printing and gunpowder… in particular did. They came from China.

Koenigsberger: Is that generally accepted?

McNeill: It was in China five centuries before it was in Mainz.

Koenigsberger: It was invented before. But did it come? And even gunpowder … ?

McNeill: Well, it seems to me, this is a real proposition that is worth exploring. We haven't got documentary evidence for it. On the other hand you know very well that there were literally thousands of people moving back and forth between China and Europe in the Mongol Age, literally thousands. Marco Polo happens to be one who wrote about it. But there were many hundreds of others who did it, too. And I always like to remind of the example of William of Rubruck[8] who went to the Mongol Court and found a woman from about 20 miles from where he'd been born in Flanders. At that court. And they talked Flemish together. Well, that kind of movement back and forth makes it—it seems to me the burden of proof rests upon whoever says that Gutenberg in Mainz didn't get the idea of printing from someone who knew something about what the Chinese were up to. And the generation of gunpowder—similarly. You probably know the first portraits of guns are in China and Europe are one year apart, 1327 and 1328. And the kind of gun that's portrayed is sort of … with a big arrow sticking out of it, is so alike. If you think that was independent, I just think it's ridiculous. It can't be independent.

Koenigsberger: I have heard Needham[9] give a lecture on [this] … In the end he remained inconclusive.

[8] Willem van Ruysbroeck, thirteenth-century Flemish Franciscan.—eds.

[9] Joseph Needham (1900–1995), British biochemist, historian and sinologist, initiator and editor of the still ongoing multi-volume *Science and Civilisation in China* (Cambridge: Cambridge University Press, 1954–).—eds.

McNeill: Well, one does not have documentary evidence. If you say it isn't true until you have documentary evidence, then, of course, it is not true. But that seems to me a naive assumption that everything that matters got written down and survived to our inspection today. There was an artisan tradition—people moving back and forth who talked to one another that didn't write. And that's how that kind of thing can move back and forth across 5000 miles. And I would say did. And the proof must rest on whoever denies the possibility. You know that the line of communication was there—the possibility of such movement. You know it was in China—in the case of gunpowder 300 years, in the case of printing 500 years—before it was in Europe. And to say Europeans invented independently because they were European, because there is no documentary evidence that says it came from China seems to me ridiculous. …

Elias: I just wanted to come back for a moment to the use of metaphors and terms like 'civilisation' in this general form. I am a bit doubtful whether it is technically useful to use disguising words like 'civilisation' in that sense for what is essentially, what are essentially human beings and which should be, in my opinion, conceptualised as such. For instance, if you speak about the coming of writing. Now we know pretty well that writing came into use very slowly. The earliest form of writing which we know of at the moment are the Sumerians. And we know pretty well that a particular social constellation was necessary, was really the constellation within which writing had a function, namely a big temple household, and that with the development of large-scale organisation means of control such as writing came into use, and for several centuries it had no literary use. It had no other use than administrative, or the Sumerian language and writing had no other use than that. And then under certain human conditions it became a medium for literature and for hymns and for other functions. So, if you want to understand any development of this kind, you have to go back to the development of the groups who, in whose midst these developments took place. If you simply say 'civilisation', then it becomes just as the word 'culture'—a purely static thing, as if at one time writing was there and now they had a 'civilisation'. I think—of course I mean I am speaking for myself—I can say that metaphors which do not express clearly the processual character, the changing character of what we deal with, distort our whole thinking about them. I would think the same thing goes for the term 'system', although the term 'system', you do it in the sense of a developing thing…

Wallerstein: My term is 'historical system'.

Elias: … but in the end it becomes something purely static and remote from the human beings with which we have to deal. So I just wanted to say—there's much more to be said—I think that, fundamentally, the science of history and the science of sociology—however you call it—will come to flower only if you very firmly relate all the various facts to groups of human beings in relation to each other. And I think that I have given, I have at least tried to say what is the social formation which is the primary subject-matter, I have called it self-ruling groups or survival groups, that is to say, groups which in one way or the other rule themselves and struggle for, are ready to struggle for their self-perpetuation and their survival over others. This, firmly should be the unit of reference for whatever we are talking about. And there, of course, Professor McNeill, I must say, there's no absolute break of continuity between simpler people and civilisation. Civilisation seems to establish a static break—there are the simpler people and there are the civilisations. I don't know of such things. I see a continuity of development between what we call 'simpler' people, and I am not sure, Professor Elwert, if you have said it, but we are very well able to determine what characteristics, what structural characteristics, distinguished the societies of simpler people from those of more complex ones. We can follow the functional differentiation in such societies very well. We can, for instance, say at what stage specialised people took over the function, the ruling function, and at which stage ruling functions are still performed by the same people who grow their food. So it is possible, with very great precision to give the various stages in the development of human groups and I think we put unnecessary barriers, if I am allowed the word, idealistic barriers before our thinking, if we use terms like 'civilisation' as if it were a thing *per se* outside the human beings with which we have to deal.

McNeill: Well, I try not to make it outside the human beings, but I must admit the word does exist and is put in the air and can be so understood apart from particular barriers of the cultural tradition that takes a particular form. The notion that it is static seems to me almost the exact opposite of what I conceive civilisation to be. In the paper I said it becomes a sort of centre of turbulence, in a certain sense a centre of social change with tends to radiate outward and involve marginal peoples, adjacent peoples in an ever-enlarging geographical expanse. Ever-enlarging is not quite right. Sporadically enlarging. So that it's not a static notion at all, but one in which intensified pace of social transformation is the leading characteristic. I don't readily surrender the term civilisation because it seems to me there are differences in social patterns, social cultures. And if you simply use the word 'culture' for them all, you confuse or you decline to make a distinction, which is a significant one—between the more complex and the less complex. Now where you draw the line is of course arbitrary. And my emphasis upon sacred aspiration as a bridge across often competing political structures, seems to me a significant thing. To say only self-governing units are the ones, this has a tangibility that makes it acceptable and things which... units of human groupings that are larger than self-governing units somehow should be dismissed, as I say, seems to me to dismiss linkages which are very real across competing states or adjacent states. And to surrender that as metaphysical I'm not ready to do, in spite of the fact that you don't like it. I still think there has to be something that links all the Chinese heirs of Confucius, as I have said, whether there's one state or a plurality of states. And that links the Europeans of the Latin world, the Latin church that has never had a single state but there is still something that I call European civilisation. And most of us really do think there's a European civilisation that isn't simply embodied in a particular state or series of states, but links. And it's largely this heritage of aspiration, the Christian heirs of Christianity, Christendom that makes European civilisation European civilisation. Transmogrified, translated, secularised, diversified, but what holds it together is that common religious past. And that is true of other civilisations as well. Perhaps I am misguided—in my world view, that is the case.

From Shamelessness to Guilt

Keith Hopkins

Laughter is the prelude to fornication
St Clement of Alexandria, *Paedagogus* 3.27

In *The Brothers Karamazov* Dostoevsky brilliantly portrayed Christ's fate when he returned to earth in the sixteenth century, on the day after a grand *auto da fé* in Spain, in which 'almost a hundred heretics had been burnt for the greater glory of God … in the presence of the king, the knights, the cardinals, the most charming ladies of court and the people of Seville'.[1] Christ by his charisma swiftly drew an admiring crowd, performed some miraculous cures and raised a young girl from the dead.

The Grand Inquisitor passing by, saw the crowd, recognised Jesus, immediately had him dragged off to prison and sentenced him to death, so that he would not disturb the social order which the Church had

[1] Fyodor Dostoevsky, *The Brothers Karamazov*, trans. David Magarshack (Harmondsworth: Penguin, 1958), chapter 5.

K. Hopkins (✉)
Cambridge, UK

A. Bogner, S. Mennell (eds.), *Civilisations, Civilising Processes and Modernity – A Debate*, Palgrave Studies on Norbert Elias,
https://doi.org/10.1007/978-3-030-80379-7_10

established. At dead of night, before the execution, the Grand Inquisitor visited Christ in his cell and upbraided him for having resisted the three great temptations in the wilderness: worldly wealth, earthly power and the use of miraculous power to escape danger. In his view, Christ had reached the wrong decisions, and in so doing had imposed a huge burden on his followers: poverty, powerlessness and the freedom of moral choice. The Catholic Church, Dostoevsky maliciously added, had been doing its best for centuries to repair this damage; Christ could not now be allowed to endanger its achievements.

Dostoevsky's fable is disconcerting, because it dramatically reaffirms how much of Christ's original teaching had been lost, when Christianity developed from a radical sect of believers into the universal religion of the established church, favoured by emperors and in alliance with the State. Jesus's scathing comment, for example, that it was easier for a camel to pass through the eye of a needle than for a rich man to enter the kingdom of God (Mark 10. 25) was rendered socially innocuous by allegorical interpretation: for example, wealth was admitted to be poison, but alms-giving, which redeems men from sin, was poison's antidote. Rich sinners could therefore redeem themselves by charity to the poor, or to the Church.[2] In this way, wealth was turned to the advantage of the Church, and of the rich. The rich could buy virtue by charity. The revolutionary, proletarian element of Christ's original message was lost in the corruption of ecclesiastical power.

But for me the most disturbing element in Dostoevsky's story is the suggestion that within the existing religious framework of revelation and redemption, Christian beliefs could have been differently conceived. The Christian Son of God should have succumbed to temptation. And then he could have been portrayed, not as infallible, punitive to sinners and forgiving to the good, but as himself fallible yet forgiven. The idea is not as far-fetched as it may sound.

In the Indian Śivas myths, which are roughly contemporary with early Christianity, the God himself and holy ascetics repeatedly succumb to temptation, often sent to them by other gods. In one story, for example, it is told that a god once practised asceticism for a thousand years; the other gods were envious and so they sent a beautiful prostitute to seduce him. Successfully. Four themes in the original tale seem important. First, a

[2] Ambrose, *de Helia et ieiunio* 76; G. E. M. de Ste Croix, *The Class Struggle in the Ancient Greek World: From the Archaic Age to the Arab Conquests* (London: Duckworth, 1981), p. 436.

prostitute who is below morality reminds the ascetic who is above morality that the Spirit needs the Flesh. Secondly, long asceticism and abstention from sexual activity fires sensuality, by making the ascetic both more attractive and more virile. Thirdly, the ascetic's repentance at his lapse fires his subsequent ascetic zeal. Finally, Śivas is god both of asceticism and of lust; he combines the opposites which in Christianity are split between God and the Devil. The Indian combination is alien to Christianity, but the underlying problem is familiar. How can we live in this world enjoying its pleasures, and yet free the spirit and acquire salvation, whether in this world or the next? I am not suggesting that early Christians consciously faced this choice in constructing their godhead. I am suggesting, as Dostoevsky did, that these problems could have been differently resolved within a religious framework. Men create their own god as an idealised image of themselves.

My problems in this paper can be simply put. Why and how was it that western culture came to be dominated by a religion which identified holiness with sexual purity, and which laid a heavy burden of guilt on the many individuals who failed to meet its exacting standards of generosity, other-worldliness and sexual continence? Why was it that sexual behaviour and sexual pleasure became the object of moral preoccupation? Why did ethical anxiety in our culture fixate on sexual guilt, rather than on say food or pollution, as in Jewish or Hindu culture? And what determines variations in the density of guilt as between groups and cultures, and changes in sexual liberation inside a culture in different periods? These are difficult questions to answer. But posing them already suggests the lines of an argument: first, there was a moral revolution in Roman society in the first few centuries AD, and secondly, Christianity was the main vector of this moral revolution. The conversion of the Roman empire to Christianity laid the foundations of moral culture in western Europe, and set a mould for moral discourse and self-examination which has never been broken.

Because of shortage of time and in spite of the well-known dangers, I shall now advance two ideal types. I shall crudely outline and contrast Roman Christian attitudes to sexual behaviour, first somewhat abstractly in the style of Foucault, and then with examples in the style of the micro-historian. These crude contrasts cannot do justice to the wide variety of beliefs actually found in pagan and in Christian circles throughout the Roman world.

In pagan Roman society, sexual behaviour was not a central concern of legal or religious prescription. There was a legal and customary indifference to infidelity by men and in some circumstances, by married women; there was a tolerance of homosexuality, roughly akin to what is found in liberal social strata nowadays. The sexual act was highly valued. The erect penis was a symbol publicly and commonly portrayed in streets and houses, on lamps and in ornaments worn around the necks of children. But there was no coherent, unified morality, authoritatively imposed. Instead there was a diversity of private ethics, which constituted a self-indulgent luxury, with which the idle rich could pad out orthodox morality. But ethical variety did not reflect a lack of ethical concern or lack of thought. Ancient ethics, in Greek and Latin, were thought out, written down, and used as an instrument of education, primarily for men, in formal academies. But there was no fixation on sexual prohibitions, and such sexual abstinence as philosophers practised or recommended was not enshrined in religious or legal codes.

Pagan ethics were concerned with the exercise of personal liberty and with the development of an individual's authority over himself in the pursuit of pleasure and personal salvation. Pagan philosophers were concerned with configurations of thought which can be located along an axis, which led at one extreme from sexual abstinence and vegetarianism to the body beautiful and to health, sometimes to homosexual relations, sometimes to marriage, and to the good life, and eventually at the other extreme to death and to life after death. This axis of thought offers some similarities and some continuities with Christianity, but the apparatus of Christian belief and of moral coercion and the contexts within which Christian ethical teaching was applied and practised were completely different.

Let us now turn to Christianity. Christian moral teaching in the fourth century, at least, fixated on sexual guilt, on the battle between *will* and *desire*. The victory of Christianity in the fourth century was accompanied by a massive increase in the social density of guilt, and by an unprecedented internalisation of moral conscience. Christian morality differed from pagan morality both in substance and in coercion. Christian morality was universally applied, among all believers, and enforced by religious education, weekly sermons, by fasting, by confession, by penance and by the heightened fear of damnation and eternal punishment. Christ died to save man from sin, but the Church encouraged an exquisite examination of the conscience, which aggravated the sense of sin. The Christian clergy promoted ideals of virginity, and of perpetual chastity, which devalued marriage and procreation. Christian ascetics denied themselves the pleasures of the flesh, and in suppressing their own desires and in preaching its suppression in others, they inevitably acknowledged sexuality's pervasive power. Their own and others' failure to achieve a perfect control of sexual thought and deed bred guilt. Guilt was the god-child of moral repression.

St Augustine reluctantly acknowledged the legitimacy of sexual desire within marriage, quoting in support God's injunction: increase and multiply. But at the same time, Augustine realised that sexual desire was a result of the Fall, when God gave man procreation as a consolation for death. The orgasm was a symptom of Adam's Fall, a symptom of God's punishment, in that the orgasm swamps the rational mind and involves a loss of rational consciousness. The detonation of orgasm within marriage was a sad concession to a couple's fallen nature. 'You descend to procreation in grief: that is the punishment of Adam. We cannot transcend that punishment'. In other words, to punish man, God created a tension between the capacity for pleasure and conscious control. Sexual lust, which burns indiscriminately, both for legitimate and for prohibited love objects, can be controlled only by perpetual moral vigilance. Sexual lust is a continuous source of danger. In Christian thought, the feelings in and surrounding the sexual act became a battleground for the conflict between will and desire, a battle between good and evil. It is not a battle which is settled by fidelity within marriage, by the simple avoidance of adultery. The battle against lust cannot be won even by perpetual chastity, because it is a battle which has to be fought in the mind. Yet ironically, by fighting the battle, Christian ascetic preachers stimulated the enemy. They could not defeat lust, except by calling everyone's attention to it. They made the prohibited sound desirable. I quote Tertullian's sympathetic account of an orgasm:

> In a single impact of both parties, the whole human frame is shaken and foams with semen. The dampness of the body is joined warmth of the soul ... In that last breaking wave of delight, do we not feel some part of our very soul go out from us?[3]

By such teaching, Christians came to recognise themselves as subject to lust, and to guilt for experiencing lust. These feelings are part of the cultural heritage which early Christianity created for us; part of our general experiences of ourselves. I am not saying that Christians created guilt, but that they thickened the density of guilt in western culture, and in doing so they changed not only moral discourse but also the perception which

[3] Tertullian, *de Anima* 28.

western men and women have of themselves and each other. In estimating the value of the change, we should balance the disadvantages of guilt, against the advantages of the introspection which guilt evokes.

Let me go back now to pagan Rome in the first century AD and illustrate Roman attitudes to sexuality. Juvenal bitingly and openly describes the self-indulgent exhibitionism, and the sexual exploits of women in court circles at Rome.[4] The women's ostentatiously competitive heterosexuality and homosexuality seem surprising even in today's allegedly 'liberal' morality. The ostentation and the competition are important because they indicate the social pressure, at least on some Romans, to live their sexual lives openly and publicly. We are not faced simply with individual acts of psychopathology. The sheer variety of corroborative evidence indicates that we are dealing instead with an established and condoned social pattern, admittedly restricted to certain social sets and openly criticised by moralists. But the rhetoric of moralists reveals a sneaking admiration for and a covert fascination with these acts of social heroism.

Let me illustrate aristocratic mores further with a comment written in the middle of the first century AD about adultery and divorce by a senator and philosopher, who was the emperor Nero's tutor:

> Is there any woman now who blushes at divorce, when distinguished and aristocratic women number their years not by consuls but by their husbands? They leave one husband in order to remarry, and marry in order to divorce. When divorce was rare, they shrank from the scandal. But now every gazette carries notices of divorce. And repeated news is an effective teacher. Adultery brings no shame, when a woman's only use for a husband is to inflame her lover. Chastity is simply proof of ugliness.[5]

Individual instances corroborate general impressions. Some Roman women used their wealth, their social power and their freedom in the emancipated pursuit of their own pleasure. The empress Messalina, for example, challenged a leading courtesan in Rome to a competition which she, the empress, won by having sexual intercourse with 25 men in a single prolonged session. I must insist: this was public not private behaviour. Mature Roman women of wealth and beauty were courted, and had freedom to choose their mates. The very invention of romantic love poetry at

[4] Juvenal, *Satura* 6.
[5] Seneca, *De Beneficiis* 1.9 and 3.16.

Rome during the last century BC reflected this freedom of choice. By convention, romantic love poetry was set in the bohemian demi-monde of courtesans, but by implication it offered a model of how men and women in love should behave towards each other. We can see the same model advocated by Ovid in his *Ars Amatoria* written at the end of the last century BC. It was designed as a gentle guide to seduction for both men and women and was devoted to their mutual erotic pleasures; it contains the earliest surviving literary celebration of the simultaneous orgasm.

This refined eroticism was probably restricted to a narrow social set. But there is good evidence to suggest that popular attitudes to sexuality in pagan Rome were strikingly different from those approved by Christianity in later periods. First, there was public and open celebration of prostitution. Once a year, for example, prostitutes came on stage in the theatre and told the audience their specialities, their fees and their addresses.[6] Secondly, men and women bathed naked together in public baths.[7] And the public baths were popularly known as places for sexual liaisons. Of course, public naked bathing occurs in other cultures, and it can reasonably be interpreted as evidence of an effectively internalised and socialised control of lust in the face of obvious temptation. We have learnt to gaze but not to touch. From the notorious reputation of the Roman baths (which later turned some Christian ascetics against washing), it seems clear that Romans regularly succumbed to temptation. But it seems more important here that at some time during the Roman moral revolution, a significant shift occurred in the boundary line drawn between appropriate private and public behaviour. Mixed sex bathing was banned. And in addition, I suspect that there was a shift away from the Roman custom of shitting collectively in public to private and solitary defecation. I cannot date this change, but I suspect that the Christian moral revolution implied a change in body image and body language.

Thirdly, Romans arranged shows of sexual intercourse in public. For example, in Apuleius' picaresque novel *The Golden Ass*, the young hero who had been transformed by a magical mistake into a donkey—appropriately endowed—was all set to have sexual intercourse with a convicted murderess in a specially constructed cage before the crowds in the amphitheatre.[8] After his act, the murderess was to be devoured by a hungry lion.

[6] Valerius Maximus, *De factis dictisque memorabilibus* 2.10.8.

[7] Clement of Alexandria, *Paedagogus* 3.27.

[8] Apuleius, *Metamorphoses*.

But the ass who had retained his human intelligence wasn't at all sure that the lion would know which to eat first, so he scarpered before his performance. As you can see, Romans joked openly about sex, something we may take for granted. But sexual jokes were soon to disappear or at least went underground for centuries.

I shall finish this section by mentioning a fresco from Pompeii. It depicts a dining-room scene. Two ropes are stretched: taut and secured at one end to the legs of a heavy dining-room table and at the other end offstage. Two dancers are balanced on the tight ropes, a man and a woman, both naked or almost. The woman wears a yellow band which supports but does not cover her naked breasts. She is leaning forward, the man is fucking her from behind, while both of them are holding in each hand a wine glass almost full of red wine, which is clearly shown to be slightly, but very slightly at an angle, off-balance. A provocative sexual image, a fantasy perhaps, but as a dining-room fresco in a rich bourgeois house, very unchristian. All the evidence taken together implies a culturally condoned eroticism which penetrated social sets well outside the emperors' court in Rome.

Let me now illustrate how radically the moral climate changed with the triumph of Christianity, and set these changes in a historical context. By the end of the third century AD, perhaps about 15 per cent of the population had converted to Christianity. But this minority included a considerable number of prosperous and educated people, especially in the eastern Mediterranean. In AD 312, the emperor Constantine had a dazzling vision or dream (our sources differ) and was converted to Christianity. The emperor's conversion was a turning point in the development of Church power; it brought an immediate end to persecutions, and a massive increase in followers and patronage. Constantine and his successors gave rich gifts to the church at Rome, favoured Christian courtiers and supported an ecumenical council of bishops, which established orthodox belief and anathematised heretics. This intervention of state authority was a crucial development. From that time onwards, Christian bishops and priests had the political, economic and social power to impose their moral views.

From the fourth century onwards, the Christian Church repeatedly harnessed the power of the Roman state to attack pagans, persecute heretics, protect orthodoxy, and to punish immorality. For example, a law was enacted against homosexuality ('we order the statutes to arise, the laws to be armed with an avenging sound, so that those persons who are guilty… may suffer exquisite punishment').[9]

A law of AD 329 set capital punishment as the penalty for a free woman who had a love affair with a slave; the slave was to be burnt alive.[10] Apostasy from Christianity was penalised in a law of AD 381,[11] and then at Milan on 24 February 391, the emperor Theodosius decreed that pagan rites were thereafter illegal. 'No person shall pollute himself with sacrificial animals … no person shall visit shrines or temples, no person shall revere images shaped by human hands, or he shall be judged guilty under divine *and* human laws'.[12] To be sure, paganism survived, and heresies flourished. But by AD 400, the great majority of the population of the Roman empire was at least nominally Christian. And so, they were either propagators of, or targets for, a new morality. The Roman moral revolution was the joint product of mass conversion and of a powerful alliance between the Christian Church and the Roman state. Sin became a crime.

Christian moral stories and the lives of the saints celebrate the virginity of Christian folk-heroes and heroines. One popular genre told how a noble bride on her wedding night, weeping, told her new husband that she was betrothed to Christ, and that if he had any respect for her and for the Lord Jesus Christ he would leave her alone and unsullied. I cannot here go through the many variations of the story which are found. Some husbands capitulated after a token struggle; some wives made a public declaration of their intended chastity at the wedding feast—to the consternation of the groom and parents. Some husbands simply refused, or as a consolation consented to dedicate their first child to life-long chastity from birth. Such dedications (*oblationes*) became a common practice among fifth-century aristocrats. To give the flavour of these fables, let me cite briefly from the life of St Melania, a rich Roman aristocrat who lived in about AD 400. She married reluctantly at age 14. She immediately asked her husband to share a life of chastity and continence together. He refused. Their first child, a daughter, was consecrated to perpetual chastity. Then, after a difficult labour, Melania produced a premature son, who died. Melania fell seriously ill. She told her husband that if he wanted her

[9] *Codex Theodosianus*, 9.7.3 [AD 342].
[10] Ibid., 9.9.1.
[11] Ibid., 16.7.1.
[12] Ibid., 16.10.10.

to live, he should give his word that he would live in chastity. He agreed. She, delighted with his declaration, immediately began to recover.

This is a story, a romance, a mythical prescription of an ideal life. It may not be all true. For my present purposes, that does not matter much. Such stories were told and retold. They were the currency of the moral culture, just as coins were the currency of the economic system. They elevated social heroes and heroines above everyday life, and at the same time provided readers and listeners with a moral map which helped each of them find his or her own place and direction within the moral order. Such stories upheld an image of what was right, of what God ideally wanted. They set a standard of moral and sexual behaviour. Judged by this standard, most people fell short. Consecrated virginity, as the bride of Christ, was clearly the ideal state. But if that was impossible, then true believers should aim at holy poverty and sexual continence within marriage. One of the ideas behind this paper was that such ambitions, once accepted, aroused guilt in those who could not match their own expectations of themselves.

The cardinal virtue celebrated in the New Testament was love. The cardinal virtue advocated by the fathers of the Church from the third century to the fifth century was chastity. Collectively they were obsessed with the fear of sexual sin (*porneia*). According to St Clement, Greek philosophers taught that one should fight off desire so as not to be subservient to it. But the Christian ideal was not to experience sexual desire at all.[13] In a strenuous effort to achieve this objective, thousands of monks from the end of the third century onwards sought solitary salvation by isolating themselves in cells in the Egyptian and Syrian deserts. There they performed almost incredible feats of competitive asceticism. St Jerome described 'with a thrill of admiration how he had seen a monk who for thirty years had lived on a small portion of barley bread and muddy water; and another who lived in a hole and never ate more than five figs for his daily meal'. Jerome commented that some might not believe all this, because they had no faith; but for those who believe, all things are possible.[14]

[13] Clement of Alexandria, *Stromateis* 3.57.

[14] W. E. H. Lecky, *History of European Morals from Augustus to Charlemagne* (London: Longmans, Green, 1869), vol II, pp. 114–15; Saint Jerome, *Vita Pauli* and *Vita Hilarionis.*

Two points should be made. First, the monks went into the desert to find their souls and were confronted with their bodies.[15] Secondly, the irony of their enterprise was that the harder they tried to suppress their sexual desire, the more it preoccupied them. They tried to control their thoughts by mortification of the flesh, but solitude and starvation induced hallucinations. They were persecuted by demons. The demons often took a tempting female form. A young monk once rushed up to St Pachomius and in a voice broken with convulsive sobs told him that a woman had entered his cell, had seduced him and then vanished into thin air. With a wild shriek, the young monk ran away, and impelled by an evil spirit, plunged into the open furnace of the public baths in the village nearby.[16] Indeed, the terror and hatred of women runs as a theme through many monkish tales, but so does suppressed fascination. A young girl once went on a pilgrimage all the way from Rome to Egypt in order to see and obtain the prayers of St Arsenius; she forced her way into his presence and implored him to grant her request, namely to remember her and pray for her. 'Remember you?', replied the indignant saint, 'it will be the prayer of my life to forget you'.[17] But women were difficult to forget. A monk encountered some nuns on a road; as soon as he saw them, he stepped off the road. But he was rebuked by their leader, who said 'If you were a perfect monk, you would not have looked at us or noticed that we were women!'[18]

The basic problem was that the conscientious believer could not win. St Anthony, the founding father of the desert hermits was, according to his biographer [Athanasius], 'a daily martyr to his conscience, always fighting the battles of the faith'[19] and the main battle was with sexual desire.[20] John Cassian, who brought the eastern monastic rules to the west, considered that monks were responsible even for their night-time thoughts. Sexual fantasies in the night and wet dreams might be the work of the devil, but they could also be symptoms of 'evil thoughts hidden in the mind which are brought to the surface in the repose of sleep. They reveal the hidden

[15] Aline Rousselle, *Porneia: On Desire and the Body in Antiquity* (Oxford: Basil Blackwell, 1988), p. 7.

[16] Palladius, *Lausiac History (Historia Lausiaca)* 9.

[17] *The Lives of the Desert Fathers*, book III, 65.

[18] *Apophthegmata Patrum, Pelagius and John* 4.62.

[19] *Apophthegmata Patrum, Anthony*, 47.

[20] Ibid., 11.

fevers of passion contracted when we are off our guard during the day'.[21] But if we can sin in our sleeping and waking minds, in conscience, by commission or omission, as well as in word and deed, then the reservoir of guilt is bottomless.

If these guilt-ridden anchorites had kept their antics to themselves, I could have passed them over briefly or in silence. But eremitic piety and sufferings created a powerful image among believers. The solitary search for personal salvation in the desert gave birth to coenobitic collective monasticism. Monastic practice and the ideal of life-long chastity for holy men and women spread to Italy and throughout Christendom. Why?

The desert hermits constituted a radical wing at the margins of orthodox Christianity. Previous sects, such as the Encratites, who had discouraged marriage and procreation among true believers, had been branded as heretics. But now circumstances were different. The eremitic movement started as an escape from persecution in the middle of the third century. But the numbers of monks first swelled exactly in the period when the persecution finally stopped and the Christian church extended its membership to become a universal and non-selective Church in alliance with the State.

The ascetic monks preserved two radical traditions, the tradition of martyrdom and of selectivity. Asceticism, the mortification of the flesh, the separation of virtue from happiness, was as [Athanasius said of St Anthony], a continuous martyrdom and fulfilled the old ideal that Christians were 'not born but made'.[22] Monks were true believers, standard bearers of the true Christianity. 'The man who has given himself to God', wrote St John Chrysostom, 'and has taken on the life of a monk … is more blessed than the crowned king sitting on his golden throne'.[23] Preachers tried to extend these ascetic expectations to their whole flock.

My point is this. There is repeatedly a tension between charismatic sects and the established leadership. The universal church absorbed the moral standards of its radical ascetic wing and retained control over it, but at a high cost. The church leaders internalised norms of chastity and of sexual morality (such as clerical celibacy) which were very difficult to achieve. They imposed these standards as ideals on their congregations, and subsequently institutionalised their members' inadequacy in the rituals of

[21] *Institiones* 6.11; cf. *Collationes* 22.6.

[22] Tertullian, *Apologia* 18.4.

[23] *Patrologia Graeca* 47.388.

confession and penance. The ascetic push within Christianity, with its concentration on chastity as a symbol of holiness, diverted moral energy away from the much more radical implications of Christ's teaching, namely the redistribution of property and power. When Christian ascetics and leaders went further, and indoctrinated their followers in weekly sermons and in monastic schools, with the belief that even thinking lustful thoughts was sinful, then Christians increased the density of sexual guilt in the moral economy. And finally, by splitting virtue from happiness, and sexuality from reproduction, ascetic teachers devalued happiness, sexuality and reproduction, and perhaps, in particular, they devalued the combination of sexuality and happiness.

Statements about the early development of Christianity and its relations with the Rome state provide a very limited explanation for the Roman moral revolution, simply because they prompt further questions. We surely need to know, for example, why so many people found Christianity persuasive and attractive, both before and after Constantine's conversion. And why did Christianity evolve as a religion which advocated sexual restraint and favoured asceticism? There were after all several powerful sects within or at the margin of Christianity, such as Valentinians and Carpocratians, which on the basis of the same religious tradition and the same holy texts, advocated sexual freedom, inside and outside marriage, for both men and women. Why did their beliefs get branded as heresies and lose out to orthodox, ascetic Christianity? And why did Christians hound heretics, or rather, why did the Roman state so energetically back Christian claims to be the one and only true religion? Each question breeds further questions; the regression seems endless. Christianity was obviously the most important ingredient in the Roman moral revolution. It must be an important part in the explanation of that revolution, but it is also part of what needs to be explained.

Above all, we must beware of treating Christianity as a single entity. In spite of its being a religion of revelation, whose practices were legitimated by reference to selected sacred texts and to the apostolic tradition, Christianity changed considerably and continuously. Its eventual shape at the end of the fourth century was the product of prolonged internal struggle and repeated adaptation. I can best illustrate this by outlining an alternative Christian theology advocated by the heretic Valentinus, who founded a Gnostic sect in the second century AD. I follow the account of his views given by the orthodox bishop Irenaeus—who glumly admitted that 'in the eyes of the ignorant, [heresy] seems truer than truth'.[24]

The first cause, which has no beginning, is a male/female pair, Profundity and Silence, who together bred a son (male) Intelligence and (female) Truth, who were the parents of the male Word and the female Life, who in turn had a son Mankind and a daughter, the Church. The first son, Intelligence, alone knew the true nature of Profundity, but was prevented from spreading his good news by his mother, Silence. The holy first eight entities then created a complex but attractive Cosmos of 22 virtues, including male/female pairs with erotic overtones, such as Deep Mingling, Undecaying Union, Self-existent Pleasure, Immovable Blending, and Only-Begotten Happiness. These virtues in turn had children such as Faith, Hope, Love and their last-born, a girl Wisdom. Now Wisdom the woman was jealous of her older sib, the male Intelligence, who knew the nature of the First Father, when she did not. Wisdom conceived an illicit passion for the First Father, and produced a son the Demiurge (or Creator), who like the fallen angel Satan in the Old Testament cosmogony, was dismissed from aetherial space and created our material world. Thus our material world was the vicarious product of Wisdom's tears, created in fear, perplexity and grief. To prevent the repetition of this illicit misadventure, Intelligence created Limitation or Boundaries, and to redeem the material world he created Christ and the Holy Spirit.

This cosmogony seems appealing, a family story in which the feelings and love of men and women are projected into the Godhead. The paradox of the good God and his imperfect world containing evil is resolved by attributing the responsibility for its creation to a lapse by Wisdom and to the Demiurge, who is a lesser god;[25] he is the God of Israel who made inflated claims for himself: 'I am a jealous God; there is no other God besides me'. But by announcing this, it is argued in the Secret Gospel of John, the Demiurge 'indicated to the angels that another God does exist; for if there were no other one, of whom would he be jealous'.[26]

The Gnostic initiates can achieve a perfect knowledge of this cosmogony. By so doing, they learn to put the Jewish Christian Demiurge or Genitor in his proper place. The initiates gain redemption, which Irenaeus typifies, as a freedom to act in any way they please. Those who have seen

[24] Irenaeus, *Adversus Haereses*, Book 1 chapter 2.

[25] Elaine H. Pagels, *The Gnostic Gospels* (New York: Vintage Books, 1979), p. 44.

[26] Quoted ibid., p. 35.

Gnostic truth do not need to obey the bishop, engage in good works or practise Chastity; they have a purer form of faith, like the faith of the apostles Philip and Thomas. Like Thomas, they have no need to seek martyrdom by confessing Christ to the Roman authorities. They who truly believe are *in* this world but not *of* it. If a man does not make love to a woman, he shall not attain truth. But those who do not believe in the Gnostic cosmogony, the orthodox Christians, are *of* this world; they will not achieve truth in any case, because they make love out of carnal desire. Therefore they should practise continence and good works.[27] In short, true believers are perfect. The orthodox Christians are contemptible and ignorant. On the whole, I share Pagels's admiration of this sect and their cosmology. Its imagery is both familial and erotic, it gives women an important position in the Godhead and by implication in the social life of believers, and it subverts the hierarchy of the church. Those highest in the church are not those closest heaven. But by the same token, just because of its irreverence for hierarchy, orthodoxy and organisation, the Gnostics lost out.

[27] Irenaeus, *Adversus haereses*, Book 1, chapter 6.

Discussion of Hopkins's Paper

Transcript

Wallerstein: I have a question. I don't know whether you feel shameless or guilty about the fact that you're unwilling to put this shift into a category of explicability, but what strikes me from your account is the rapidity of the shift is quite striking. Although you didn't recount the degree to which the shift ... the length of the period over which shamelessness was in fact the norm in itself might be of some interest. But it does raise questions actually about the whole idea of moving towards moral control because it isn't a slow movement, it's a sudden movement. It then lasts for an awful long time in one way or another, the guilt. And now if you think the twentieth century where one could say there's been a rather dramatic shift from guilt to shamelessness. I don't know whether we've quite reached Roman heights yet but I mean there has been a steady liberalisation of sexual mores for 75 or 100 years which is perhaps as dramatically swift as the other, which might lead you to a kind of cyclical theory although it's a very long cycle. But in any case, I must confess that I feel that everything is explicable—even if complicated—and I wouldn't accept reserving certain areas to the mysteries of chance. And I urge you to reflect upon whether, if this shift from shamelessness to guilt confirms our

Transcript (✉)
Berlin, Germany
e-mail: Artur.Bogner@web.de

A. Bogner, S. Mennell (eds.), *Civilisations, Civilising Processes and Modernity – A Debate*, Palgrave Studies on Norbert Elias,
https://doi.org/10.1007/978-3-030-80379-7_11

theories, that in the present twentieth century one from guilt to shamelessness disconfirms them.

Burguière: I want only to ask one question to Professor Hopkins … of this very interesting paper. Would it be possible perhaps to try to explain this moral revolution more by a pre-Christian revolution than by a Christian revolution? In other words would it be possible to try to follow the Elias way of explanation and to try to explain this change by a new configuration of power, I mean in the relations between the state and society and precisely … You have pointed out the influence of the moral crisis of the third century by the increase of the sense of grievance, guilt and so on. But would it be possible, perhaps, to see in this change, in this new morality among the Roman elite, the non-Christian Roman elite, not only toward sexuality but perhaps toward family life, toward the morality, toward the conjugal relation for instance, a new model of morality and of behaviour toward the power. I mean … I refer here to … some explanations as there is of Paul Veyne trying to see in these new attitudes the fact that these people of elite, of the Senate and so on, were before, as chiefs of clans, leaders who had to express their male dominance, superiority in all fields, in, also in sexuality, or more in heterosexuality, became more and more, tried to be accepted by the emperor as a new kind of concentration, of monopolisation of power as good servants, perhaps good high public servants, but… and produced a new ethic more eligible with a new conformism.[1] And also even in the Roman imperial attitudes in this public morality is it possible to relate this new, this change with some laws of taxes edicted I think, since the beginning of the Empire against the unmarried people, for instance—a sort of valorisation of conjugality which was perhaps something new. I mean this is certainly only one way of explanation. The other is this new paradigm which comes from the ascetic vein of Christianity, as you said. But the two aspects seem to be necessary to explain the change in society. And the last thing I want to say is if we look from the beginning of Christianisation of the Roman Empire till the twentieth century, the Christian paradigm has given a permanent frame of values. …

Elias: There are a few things I would like to say. First, it so happens that I have also gone a little into the problem of the period about which Keith

[1] Paul Veyne (1930–), French archaeologist and historian of Ancient Rome, allied in the 1970s with Michel Foucault. Burguière's reference *may* be to Veyne's book *Le Pain et le cirque* (Paris: Le Seuil, 1976).—eds.

Hopkins has been speaking in connection with the problem of a dissolution of a state. That was the problem really, which interests me—how... what happens if a society which is highly organised, when this high organisation begins to dissolve. So, I am delighted with what Professor Hopkins has done and said, because I always felt that my very small capability to acquire the necessary knowledge is not enough and this is very helpful to the few dips into the problem which I could take. There are a few questions which I perhaps can formulate in connection with that. First of all, I very much agree with what Professor Hopkins said. One cannot predict Constantine's conversion to Christianity—though one can explain his transfer of the capital to Constantinople. That, I think, can be explained in a variety of terms. One cannot explain this conversion to Christianity, but one can say that it had a very good function for the state to have also the now not inconsiderable organisation of the Church at the disposal of the control of the State. It now allowed the emperor to control not only through lay, military means the whole empire, but now to control it also through the organisation of the Church. So I think one has to balance lack of predictability, individual initiative, which exists, with the function which it may have. I have also been puzzled by the enormous turn towards asceticism, by the extent to which the Roman mythology was completely put aside by the new mythology. I do not use this in a derogatory sense at all. There was, apart from the central tenets of Christianity, a lot of mythology, new mythology which really came in. I will give one example. I remember Salvianus in the fourth century, I should think—I'm not quite sure—writing a book, under the name of 'Timothy', on the Church, in which he advocates that people should leave their money more to the Church, because the Church needs the money and in which he then threatens the people.[2] They should consider the rich people, what would happen to them after death if they don't give their money to the Church, which of course will absolve them from hell, but so he describes what will happen to them in hell in words which are really very odd. He describes if you are then in hell, and you will burn and you will have thirst, you will cry out, 'Father Abraham, send Lazarus to me with a wet towel to give water to my tongue'. So there's a new mythology. There is Father Abraham very oddly and Lazarus as the symbol of the servant, of course, no high-powered man like Father Abraham could do it himself. There would

[2] Salvianus, *Ad ecclesiam* (also known as *Contra avaritiam*); Salvianus in fact lived in the *fifth* century.—eds.

perhaps have to be a slave or someone low down and Lazarus had become the symbol of the low man. And it is by no means the only mentioning of Father Abraham, as some part of the new mythology. There is a question with regard to asceticism in the very connection with regard to Salvianus. He himself was a married man and then he one day decided to go into a cloister, into a monastery. And his wife went into another monastery, a neighbouring monastery. And they went for a time into the monastery.

And I've just been wondering to what extent this attitude at that time had already become part of the career structure of a person who wanted to make his career within the Church. So this is one question. There is a mixture between the undoubted turn towards asceticism which became the normal standard within the Church and … all those who belonged to it. And I think one can, or one may perhaps, if I may make a suggestion, distinguish between two periods: the period in which outwardly the West Roman Empire or the Roman Empire as a whole was still intact. I mean the period of Constantine and his immediate successors, in which the Church became a real state institution. And of course, one will have to say that it is very doubtful whether Christianity could have played the role which it did play if it had not become the state religion of Rome. This period of the state religion of Rome was a period which was very important for the formation of Christianity itself as well as for its role as transmitter, transmission machinery of knowledge from antiquity. When the state dissolved then the Church was the only large-scale organisation which remained reasonably intact. And this role of transmission of knowledge has been tremendously important. But there's one last question which I wanted to ask. The turn to sexual asceticism was not confined to Christians. If one looks at Julian the Apostate, then one finds that he too had not only considerable dedication to magic, but also he was married, but he was with emphasis on fidelity to his wife, as far as I remember.[3] That is to say he was appalled by, reacted strongly against, the loose morality of the previous centuries in imperial Rome. My question is really this: To what extent was this asceticism, this what you call the moral revolution connected with the, on the one hand, in the first period with the fact that now a sphere of life was brought under state control, which formerly the state did not control? I mean in the form of the Church, the state was able to control spheres of life which it formerly did not, could not control. The

[3] Flavius Claudius Julianus, Roman emperor from 361 to 363, a learned man who was the last non-Christian emperor.—eds.

second problem is, which comes nearer to the problem, why the old secular philosophies completely disappeared. I mean this is also, I think, a problem one has to take in account. In the late Republic, you still have people, the great families of Rome, imbued by Stoa and other philosophical ideas of a secularised kind. A secular morality, which was not the morality of a sort of wild sexual indulgence, but there was a real sense of stoic morality and also the Epicureans had a fairly wide following. We then see, I mean, what I mean is one should not perhaps paint the counter-picture to the moral revolution entirely in terms of these indulgences. There was also a morality represented by Cicero, by Stoics, by Epicure and other people which had very clear standards in the late Republic, and also still lasting in, a little, into the first century probably of the imperial time. The fact that pagans also inclined to stronger control of sexual tendencies, has it anything to do with the growing insecurity of life? Which is very remarkable, the whole disappearance of the secular philosophy. And the whole transformation even of the pagans into a new sort of religion, going away from Jupiter and Zeus and going more ... I think the godhead of Julianus, if I remember rightly, was the sun-god.[4] So I mean there is a transformation of religious beliefs which goes beyond Christianity. And perhaps, in trying to explore why it happened in Christianity, one would have to take account of these two facts: the disappearance of secular philosophy and the increase in a new bent towards, even in part monotheistic, beliefs even among pagans.

Hopkins: I think rather than deal with each person in particular what I'll do is take up various points that seem to me interesting where I can say something, which I find interesting, about it. What I've done is I've presented a case study and I think it will be a great pity if the discussion concentrated exclusively on the details of the case study. What's interesting about it, it seems to me in terms of general theory, is that Western society has gone through a number of oscillations of sexual liberty to sexual restraint. I myself think that the oscillations like that are within a general developmental pattern or directional pattern, towards greater self-control. I take your paradox. I thought it was very interesting, a (well-set) question, as it were, but I don't see a real opposition, in spite of Dostoevsky. As I see it, what happened is that the Church, the institutional strength of

[4] In this not very clear sentence, Elias is alluding to Julianus's Fourth Oration, the *Oration to Helios*, in which he made Helios the main deity in his attempt to revive pagan religion.—eds.

the Church, through the practices of penance, confession, sermons, and teaching, and celibacy as an ideal …

McNeill: When do you think penance and confession?

Hopkins: Fifth, sixth century.

McNeill: It came from the Irish.

Hopkins: A pleasantly dogmatic assertion but it may not work.

McNeill: In the ninth century.

Wallerstein: Penance? That's penance that you're talking about?

Hopkins: … which have earlier components than the ninth century. The Penitentials; and I can, you know, that's a technical argument, but I mean I'm sure I can find much earlier public confession—I can find fifth century …

McNeill: Confessions, yes. But not penance.

Hopkins: If you look at the… what would I show you. I mean … sources of …

McNeill: I was told by my father. Therefore I know that.

Hopkins: If you look … If you want to play this game I would say, do you know the *Apophthegmata Patrum*? And I suspect you don't.

McNeill: No.

Hopkins: Right. If you look at the advice given to the fathers in the desert which is called the *Apophthegmata Patrum*, which are fourth and fifth century concoctions which survived in Ethiopian and Syriac form, then you will see both penance and confession. And, you know, I mean I can play that scholarly game with equal dogmatism and I suspect, in my own field, more learning. So, if, in other words, it depends. The spread of it in Western Church is much later. But as a …

McNeill: It's the Latin Church you're talking about. The Latin Church.

Hopkins: Yes, as a common practice. But actually as an institutional form, you can show it certainly early, late fourth century, early fifth century. But look, what I'm trying to say is that the Church acts as an ossifier and a rigidifier of a practice, which then is fed back into a moral norm which is internalised in the population. Obviously, the degree which it penetrates is highly diversified. And it hits different people with different density. My general argument was that both that the aggregate and the spread of guilt was greater in the fourth and fifth century than even before. And that, as it were, creates a pattern from which Western culture has never broken out—the waves in the density in which it's been applied, Professor Wallerstein, so therefore that the Church and the person, the individual, are, as it were, in dialectic with each other. We heard yesterday of Professor Wallerstein's intellectual imperialism and I'm not at all surprised to hear that he believes that everything can be explained, even with difficulty. I just belong to another camp and say that I don't think everything can be explained and I suppose, to borrow a phrase from Professor McNeill, the burden of proof lies with you. I mean, if you can explain it, I shall listen.

Wallerstein: I'm not sufficiently learned.

Hopkins: I think it's a very interesting question within sociology as to whether everything can be explained or whether there are some things which can't be explained. About the relative freedom of nowadays and whether there's been a unified direction. I remember Professor Elias a long time ago making the point that we now have freedom, which appears to be freedom but only within the limits of strict social self-control. In other words, apparent freedom is possible, sexual freedom is possible, because our internalised control is now so deeply imbedded. Say, naked bathing on the beach can occur, simply because we are controlled, self-controlled to an extent that say 500 years ago would have created much different behaviour, riot, rape, or whatever.[5] So, in other words, I don't find obvious exercise of sexual liberty as necessary in negating the argument that we now have greater self-control.

Wallerstein: Well, there's certainly a good deal of practice in recent days that would meet with the moral reprobation of a good number of people.

Hopkins: Sure. Let me go back to the argument that you produced. I mean, how many are the factors one will have to take into account, and Norbert Elias's point that in late Republic and early first century AD there were powerful pagan philosophers, philosophies, which advocated the same norms that Christians advocated. I agree completely and I pursued a rather dangerous tactic of a dramatic opposition between first century and fourth century. I do it with full consciousness of some of the difficulties about the first century. There was a wide variety of prescription and behaviour. I would like to use a concept of social heroism. By social heroism I mean behaviour which is highly admired, but which is not necessarily universally applied. And what I'd like to argue is that in the first century AD social heroism was, in some circles, associated with sexual liberty even though there were countervailing (a) practice fidelity and marriage and

[5] See Norbert Elias, *On the Process of Civilisation: Sociogenetic and Psychogenetic Investigations* (Dublin: UCD Press, 2012 [Collected Works, vol. 3]), p. 182.

(b) philosophies which were close to ascetic, Christianity among them. In the fourth–fifth century there was a monopoly, by the Church in effect, advocating sexual asceticism and social heroes were monks. And so that I think it's that combination. I can't, alas the evidence doesn't allow it, say how behaviour has changed in the fourth and fifth century because most of the surviving sources are prescriptive, are sermons or letters, and they're not descriptions of behaviour. So what you've got is in the first century I can describe behaviour. In the fourth century, I can only deal with prescription. But I think, what I would like [in the] discussion, if I could ask for discussion, is how, is it likely that the density of moral guilt changes within a civilisation over time and how far does the cultural construction of personality, and of superego, change within our civilisation, within our culture over time. Most anthropologists take human nature for granted. Most sociologists resist psychologistic explanations. The first seems to me improbable. The second, it seems a pity that we so much resist looking at human personality as itself a cultural construct. This was a case study, an adventurous case study, in an analysis of cultural creation of personality.

Wallerstein: But that's an explanatory model. And why are you saying that you can't explain it? If personality is a cultural construct, which I absolutely believe, then one has to explain why it gets constructed in A way rather than B way at specific points in historical time and not say it's inexplicable.

Mennell (in the chair): Well, my own hunch is that is precisely what Norbert Elias is trying to do and what Keith was trying to do. I think it was a small area of inexplicability that Keith was pointing to.

Dunning: I'll be very brief. I just want to ask Keith if he would accept what seems to me to be—that there is an implicit patriarchalism in his conceptual apparatus, the kind of questions he answers. I wonder if his scholarship allows him to say, to tell us whether the sources provide us any clues to the changing balance of power between the sexes in this period. It cries out, it seems to me, for something to be said on that issue. It's so crucial.

Bertelli: Two points only. First of all, I wonder if we can be allowed to speak only of sexuality and sexual revolution, since this is only part of a more general theme. In the contact between the Hebrew society, not Christian, but Hebrew society and, I underline this, the Roman one. But let me explain. Also, the Romans had the taboo of virginity and they divided the virgins and matrons also physically with the cutting of hair and so on. I wonder if we can be allowed again to use rightly Messalina and the Golden Ass as a proof of an immorality of the Romans. It is enough to remind you of Caesar's wife. How fidelity was important in the Roman Empire, at the beginning, at least, of the Roman Empire. And how fidelity and how virginity, and that the word 'virginity', was present in the Roman Empire. So in my impression it is not sexual revolution, ... but a different sense of the human body. And to be better understood, I can remind you of the essay on holiness written by Peter Brown in which he proved how shocked the Romans were to see the use of corpse.[6] That the Christians, the Hebrew and then that the Christians had. How they used especially the martyr's corpse and the use of relics that they had. And so the sense, the deep sense of change in the use of the human body between Roman and Hebrew society, Oriental society. You must remember that the Christians survived from the Orphean mysteries and they were not certainly inside the Roman ... This is one of my observations. Secondly, and I finish, I wonder if we can really speak of a development of a pattern in one sense of self-constraint. Starting from one (hypothetic) point that it is untrue because we don't have this point, this start point. It is not demonstrated. And it will arrive to what? Because if we go on and we must discover a moment of self-constraint, we must arrive at the Victorian Era. But we have jumped the moment of the Renaissance, in which again a new different sense of the human body was discovered. But not only, but we disjoint our own present time after, I can say, 1968 when a new again different, not last, but different use of the human body has been imposed, at least in our Western society. But with another consideration connected with the first, that is the taboo of the corpse we have in our society. It is quite different to the use of the corpse again, but certainly, the use of the corpse that, for example, was so heavily present in every obituary and in every funeral of the counter-reformation era or in the saints of the counter-reformation era.

Goudsblom: A few questions, also brief. One refers to a point that has already been raised, but I don't think Keith Hopkins has gone very deeply into it yet. That is the role of the philosophical schools, the pagan philosophical schools, which, for centuries already, had been propagating forms of self-restraint and they were directed primarily to the upper classes. And I think one can explain a lot in these matters—one can explain why this advice was given to the upper classes to restrain themselves, just from reading the text. One hardly even needs to reconstruct the social context in

[6] Peter Brown, 'The rise and function of the holy man in Late Antiquity', *Journal of Roman Studies*, 61 (1971), pp. 80–101.—eds.

which the texts were written. If you, for instance, read [inaudible], advising his readers not to eat and drink too much and not to indulge in a fit of anger against their slaves, he has very good reasons to do that because there are no external restraints for a rich slave owner in a slave society to eat and drink as much as he can and to let himself go against his slaves and perhaps [he would] even hurt himself in beating his slave. So one is well advised to exert self-restraint, and this tradition—which you can find among the Stoics, Epicureans and other philosophical schools, which has been described very well I think by Adkins as the tradition of the quiet virtues against the older, loud virtues—is something which I think has entered very clearly into the Christian tradition.[7] And the traditional way—traditional in the way in which Nietzsche for instance wrote about Christianity as a slave morality—is I think one-sided. There is also a gentleman morality, the gentle morality of the philosophical schools which entered into the Christian ethic. And I would like, first of all, to have your expert's opinion, on this general idea.

Then a second question, which is partly empirical and partly also of a more theoretical nature, is this: am I right in my impression that, especially in the first and second century, the Church drew a great many women from the patriciate among its adherents? And that … there are even a few cases of the same saintly persons who advocated chastity and abstinence, managed to surround themselves with, let us say, virgins, from the higher circles, who vowed oaths of chastity and who donated great amounts of money to the early Church, thereby helping the Church to form a stronger organisation? And does this have something to do with a question that has been raised before but I don't think you have yet had the opportunity to answer it sufficiently: does it have to do with the changing balance of power, perhaps, between men and women, especially among the upper classes in Roman society?

And that, then, leads to a third question that I would like to pose which also has already been raised, about the power resources of the leaders of the Church. The little knowledge that I have, of the third and fourth centuries, especially of the fourth century, has often given me the idea that the bishops at that time acted like the ayatollahs in Iran do now. I mean, there's a remarkable similarity in a rising clerical elite—priests vying with kings so to speak as was observed yesterday—that is vying with the pagan

[7] A. W. H. Adkins, *Merit and Responsibility: A Study in Greek Values* (Oxford: Oxford University Press, 1960).—eds.

worldly authorities. One wonders what the power resources were that, for instance, made it possible for Ambrosius to win every contest that he waged with the emperor. What were the power resources and where does this whole issue of sexual abstinence come in? Is there a relation? I would guess there would be, but it would be very difficult to substantiate, I realise. So these are the three questions: the philosophical schools, the women, and the power resources of the bishops.

Hopkins: Philosophical schools: the most powerful philosophical school was the Stoic in the Greek and Roman world and, to some extent, the culture was unified. They did preach attitudes towards self-restraint, towards sexual abstinence, towards consideration—on consideration towards women, which in some ways were, perhaps, more advanced and more like our views than Christians'. And that tradition, (a) was quite old, and (b) went on into the second–third century AD, then got somewhat transformed. What's different about it from Christianity is that it, and I thought I'd said that, is that it's a philosophy which is aimed at single individuals, which has no mechanisms of enforcement. And so, although there are continuities and similarities with Christianity, it is not institutionalised as Christianity was, nor does it have the panoply of state power to enhance it, which makes Christianity such a powerful mixture. So, there is overlap with Christianity, and Stoics were powerful in philosophical circles and in court circles at the same time as I was claiming the moral climate was one of sexual libertarianism. One of you said 'How can I possibly take these illustrations as testimony of widespread sexual libertinage in Rome?' And, of course, were I taking only these two illustrations as convincing proof, you would have a good case. But I don't think I did that. I think I said that these were several illustrations among many with some …

Bertelli: There are many others that are opposite.

Hopkins: Yes. And then, since we don't have statistics of crime or statistics of sin, you have to take a view on where the balance lies. In my view, there was, within certain circles, and beyond court circles, an attitude towards sexuality in first-century Rome as witnessed by the Golden Ass or stories that one occasionally hears, by wall paintings in Pompeii, by frescoes, by decorations on lamps, which is completely different from anything in the fourth century. Now, I may not be right, but I have more than a couple of stories. … I am not saying that my view is universal. But there would be several people who would agree with me, others who would disagree. There's room for disagreement. But I think there's good evidence for my view, good evidence, cumulative evidence for a dramatic change in dominant prescriptions. Now, as for relations between men and women and how they changed, it's clear that they did change. It's not at all clear in what way they changed. And there again most of the sources, as everyone must know, were written by men and not much information is given about women. There does seem to me to be a shift in the degree to which women in the first century could control property. Very largely, they could. They could arrange their own divorces, they did control property and, because they controlled property, in my view, they were courted. And the rise of love poetry, in my view, has much to do with the degree to which women had independent access to wealth. A sort of a vulgar economic causative sequence of which I'm slightly ashamed. In the fourth century, it seems to me that there were considerable limitations on the degree to which women could control property. There were limitations on the degree to which they could initiate divorce and they were surrounded in the upper classes by eunuchs rather than by full males, slaves, which restricted their sexual access to the sort of intercourse to which they had previously had access in the first century. The only way in which they could maintain control over property—or, no, not the only way, one of the major ways—in which they could maintain control over property or gain great social status was by becoming the abbess of a nunnery. In other words, the opportunities for social status, with the birth of Christianity, were directed towards that. The mass conversion. I reckoned once with a statistic which isn't too well based, that about 15 per cent of the Roman aristocratic women converted to permanent chastity within the space of 30 years. But that is in the middle of the fourth century, not in the first two centuries, about 330 onwards. So there is a huge change in the social status or the social setting within which women typically act. But I think I can't at this moment say much more—I mean to give a lecture on the status of Roman women (a) in the first century and (b) in the fourth century. I do think there's a dramatic change. And I should say, and I take your point completely, that in the first century there were circles in which respectability within marriage … not virginity so much. Romans protected the virginity of their females by marrying them off between 12 and 14. But they did symbolise virginity as in Vestal Virgins. It was symbolically very important. But once virginity had been lost, in other words post-marriage, women who had property could divorce their husbands by their own say so, because they could change, if their tutor stood in their way, their guardian stood in their way, they could change their tutor. The conventional way of understanding these problems is to look at what I call a genetic explanation, to look at the nature of Christianity and try and see the degree to which its specific characteristics are a function of its sources. So people take Roman philosophy in the first century, Greek philosophy in the previous centuries and Jewish traditions and Christians' own particular version of them, and they see the nature of Christianity as being a mix of those. They tend to do it by thinking of Christianity as a single entity, which I'm on the whole against. And I should have said, in my view, the specific form of Christianity with which we got stuck in Western culture was very much a fourth-century creation. In my view the Christian organisation got ossified at the moment in which it had to absorb the radical wing of ascetic monks. In other words, it was caught in a highly temporary, ideological stance and it was then ossified as it was—rather as there is a Stinchcombe theory of organisational norms.[8] The idea is that if you look at the railways, the post office, steel industries, telegraph systems, aircraft industry, computing industry, that each—founded, say, at 30-year intervals—incorporates the norms which were dominant at the time of its first institutionalisation. And it's remarkably difficult, even in the post-industrial world, for these apparently modern institutions to change from their original institutionalising norms. I don't know how transferable that idea is to Christianity, but I find it quite attractive that Christianity got stuck in values which are highly specific to the beginning of the fourth century. And that it became rigidified in that stance. And it was that stance which then affected all subsequent relations. It doesn't fix all subsequent relations of church to believers, but all subsequent believers had to deal with a church which had got stuck with those particular attitudes. That's why I'm against the genetic arguments which see Christianity merely as a pot-pourri, a combination of its sources. So, I like to see it as much more adaptive, fluid, up to the fourth century, than completely the victim of its sources. Have I [answered your questions]?

Goudsblom: Well, I had a question about the power resources of the bishops and what they have to do, perhaps, with the creed of abstinence. But it's a very difficult question I realise.

Hopkins: Bishops in the east and west are slightly different. Bishops in the east are preponderantly ex-monks. Bishops in the west are quite often married men, who renounce marriage in order to become bishops. In other words, it was quite accepted that lower-order clerics could marry

[8] A. L. Stinchcombe, 'Social structure and organizations', in J. G. March (ed.), *Handbook of Organizations* (Chicago: Rand McNally, 1965), pp. 142–93.—eds.

and have children, securing their biological succession. And then when they've had children they're promoted to bishops, live apart from their wives. The wives go into a nunnery. Sometimes there are marvellous stories of the wife beating against the church door, saying 'Why have you locked me out?' But the practice varies enormously. Yet there was certainly the ideal that the bishop should be, in effect, celibate. Bishops control resources much more in the fifth century than in the fourth century. Whether in the fifth century with the breakdown of central political control they become, as it were, mini-lords, and then—at that time particularly—local aristocrats in Gaul in particular become bishops and wield local power. The power of Ambrose against Theodosius, who was made to kneel, crawl on his knees the length of Milan Cathedral.[9] A remarkable episode, I think, that should obviously be explained in terms of the overarching strength of the Church, the bishop against the emperor. But also, and I'm sorry to slip back into a highly individualistic explanation, in the magnificent personality of Ambrose. Theodosius was no slouch and why he crawled the length of that church I don't know, but he had ordered the massacre of a Holiness and population of Thessalonica and perhaps he just felt guilty.

Elias: Just quite briefly. I think the two views are not incompatible. I would like to say how much I enjoy the fact that a question is taken up here, which I think extremely important and which I always felt should have been taken up long ago. But one need not, I think, construe an opposition between an explanation in terms of the actual social situation and in terms of the perpetuation of an institution which then becomes almost self-perpetuating. And that is, I think, if I understand you rightly, what you say. But in order to test one's case I think a little more attention has to be given to something of which I am at the moment completely ignorant, that is the development in the Eastern Church. One would have to compare, I do not know to what extent the asceticism persisted in the Eastern Church, [and whether it was] to the same extent as in the Western Church. And that might be a test case for the problem of self-perpetuation.

Hopkins: The difficulty was that in the East the Eastern monks—I rely here largely, to some extent on conventional accounts—the Eastern monks were (a) much more likely to remain as eremites, in other words as solitary hermits, rather than in collective communities. And secondly, the

[9] On 25 December, AD 390.—eds.

communities in the East were, in spite of the rules of Basil, much more likely to be concerned with the saving of their individual souls, and dependent upon the local communities, than self-supporting through their works. And, of course, the difficulty about comparing Eastern and Western developments is that the Eastern developments were somewhat cut short in Egypt and Syria by Islamic expansion. There were a few oddities that survived. And about the later history of Eastern monasticism, I'm afraid I know regrettably little.

Elias: I don't think it is just a question of monasticism. It is also a question—there is a struggle between the localised control of some particular saint or picture and there are the constant iconoclastic movements in which the emperors tried to take control by breaking the pictures. But the emperors themselves, if I remember rightly and my knowledge is very limited, the Leos themselves had not a particularly strong ascetic tendency. And it's not only the monastic people one has to consider. One has to consider the whole development of Christianity in the East.

Hopkins: Yes. I mean there are some marvellous instances of individual holy men and I think you mentioned Peter Brown's study of holy men which, of course, is fundamental as well as his study of the [saint as exemplar].[10] I didn't talk at all about the body, which I should have done. Holy men really are extraordinarily important in the Eastern development. They're sort of saints. A typical case is the man who stood forty years on top of a pillar. The Stylites, Simeon the Stylite near Antioch, Daniel the Stylite near Constantinople. They stood, each of them, 30–40 years, day and night, on top of a pillar, occasionally freezing—though they were held up by a rail until, say, the flesh on their feet rotted. But one of them came down, once, in order to turn an emperor from heresy to orthodoxy, with a huge following of people. And, although he was an ignorant Syrian who didn't know Greek, the archbishop and the emperor both knelt at his feet. And he told them not to quarrel with each other and the emperor wouldn't be allowed to leave the cathedral until he recanted his heresy. And the emperor did recant his heresy. And what's interesting to me about that is the degree to which the holy ascetic acts as the social hero of the culture and can enforce his morality against both the established church and the political authority. And I think that such behaviour, which you can see replicated at the village level, as well as the power level, is quite unlikely,

[10]Peter Brown, 'The Saint as Exemplar in Late Antiquity', *Representations*, 2 (1983), pp. 1–25.—eds.

that didn't happen in the West. At least I don't think that was happening in the Latin Church.

The body. The body is now a very fashionable subject of study and I'm very sympathetic to the idea that one should examine how people's perceptions of their body and how the body, both alive and dead, is used differently within our culture, of course. And I mentioned how I thought that the body image and body language changed in the course of this revolution, public as against private shitting, public as against private intercourse. On the other hand I don't think that it's likely any more mileage is to be gained from such a single factorial analysis. I mean, you might quite rightly say that I have concentrated too narrowly on sexuality, and other things should be taken into account. ...

Elwert: Hopkins's position that not everything can be explained seems empirically valid. But Immanuel Wallerstein's position that everything can be explained seems to be the epistemological basis of sociological fertility. So please allow me to have a position which is not so close to chastity and point to some elements of explanation which may be known to you, probably are known to you which maybe are a confrontation of elements which might better apply to Khomeinis of our period than to bishops in the third and fourth century. Libertinage and divorce are not only a pleasure. They can contribute to the instability of the immediate personal environment of human beings. This immediate personal environment is the coordinate system which contributes crucially to the definition of identity of a person. There are other factors which can destabilise this environment. One, very important in our time, is unemployment, studied by Lazarsfeld among others.[11] The instability is likely to produce reactions. And I think there's a characteristic that these reactions have a tendency to reduce the range of freedom through restraints and/or that they are reactions which stabilise identity virtually through ideology. To be short, the question is whether this moral revolution you have described is such a case of constraint and ideology, elaborated in reaction to instability of the immediate personal environment.

Hopkins: That's a beautiful question—or comment. And it should be really treated as a comment. I have tried to consider that, though I didn't mention it before. I obviously wondered—like other people, say Stone in

[11] Marie Jahoda, Paul F. Lazarsfeld and Hans Zeisel, *Marienthal: The Sociography of an Unemployed Community* (London: Tavistock, 1972 [1935]).—eds.

his history of love and sex and marriage in English society[12]—whether there are, in English society at least, century-long fluctuations between sexual libertinage at one extreme and, say, prudery and constraint at the other extreme. And in some way, it seems obvious that there's a homeostatic mechanism within society in which ideology may be the fulcrum through which the change occurs. My reluctance to put that forward, without having thought through it more, is because the idea of the self-balancing mechanism within social change is one which has so often been put forward and so often been laughed out of court that I didn't want to put it forward again without thinking of how I could justify it. It may obviously be that instability, which is upsetting, creates its own reaction, on the other hand …

Elwert: Would you mind being more specific by pointing to identity?

Hopkins: Well, the difficulty is that there are so many causes of instability in a pre-industrial society that it's very difficult, unless you have much better evidence than I have, to be able to point to a particular period which is likely to have a greater density of instability, which creates a greater need to change the way in which people organise their lives. In other words, it's, well, like the concept 'crisis'. As far as I'm concerned in demographic terms or in terms of the personal organisation of life, given the way in which, say, married couples are battered by death, their lives are always unstable. And I can't see that they were more unstable. They were more unstable in the middle of the third century because of barbarian invasions. But why instability in that way should make people change their sense of identity and use ideology towards a moral change in a specific direction defeats me. I can know that it happens.

Elias: There is one very clear distinction of this period. Gradually the central authority of the state breaks down. This creates a degree of insecurity which is hardly comparable with any where the state central authority is stable and effective. So we have really a transition from the breakdown. I mean words like 'political' try to disguise from us the simple institutional change which occurred from the fourth century on, or even earlier, that the central authority became ineffective. The effective power of protection against physical injury went to local strongmen, and that created a degree of insecurity where towns were raped and ravaged four times perhaps in a

[12] Lawrence Stone, *The Family, Sex and Marriage in England, 1500–1800* (New York: Harper & Row, 1977).—eds.

year by barbarians. A degree of insecurity which has not had its equal since the centralisation process took place again.

Hopkins: Alas, if only things fitted just that nicely. The difficulty is that there's chaos in the third century. But the Roman state, although it provides general peace and order, doesn't penetrate down to the village level, the level of most people's lives, with sufficient force.

Elias: But the troops protected them for a time. And then, at certain times, one can see very clearly how the simple physical protection, and also some provision of food, no longer comes over long distances, but has to be provided locally and the local man has to provide the physical protection against invasion by others.

Hopkins: I agree, Norbert, but this is true of a 20-year period in the middle of the third century, maybe 30 years. And it's true again from 378 in certain sectors of the Empire and say from 405 afterwards. The difficulty is those dates don't fit with the period of the growth of virginity and chastity as an ideal.

Wallerstein: What's the lack of fit? Do they come earlier or later?

Hopkins: They come in the middle.

Goudsblom: Just one quick remark. The simultaneous effect of the sack of Rome and the major book of the Christian apologetics, *The City of God*. I mean, that is not fortuitous.

Hopkins: No. And the *Apologia* is a direct result of the sack of Rome.[13]

Goudsblom: Yes, precisely.

Stauth: My question was partly resolved by the discussion about the separation between Eastern and Western church. However, it remains the fact that this took place after the revolution you're talking about. And there, I wanted to point to the continuity of asceticism in Islam. And, obviously the discontinuity of the moral revolution which you were talking about, that means the high realms of the apposite time for example are very well known for their libertinage and lots of poetry is known in the West to point to this fact. So that, how does this fit to your picture of a set of institutionalising norms, which sort of as a take-off determine future developments? There's a second point. ... I don't know, but what one reads about the height of the Middle Ages is that there is libertinage, obviously, which is a sort of a moral revolution to its reverse. So, how does this fit into your configuration setting of that time?

[13] Saint Augustine, *Concerning the City of God against the Pagans* (London: Penguin, 1972); *Confessions of a Sinner* (Harmondsworth: Penguin, 1961).—eds.

Hopkins: I can only reply as a sociologist and not as a historian because you know my knowledge stops by about AD 600.

Wallerstein: Now, you're free...

Hopkins: I suppose what I was going to, my thought is that, within Western Europe, all subsequent libertinage has to cope with the moral discourse of sin which is guilt evoking. Sure, you can be free. But in order to be free (a) you have to control yourself and (b) you have to overcome the guilt which is our cultural heritage. How people in Islam do it, I don't know, but I suspect it may be easier. But that may be 'paradise out there'.

Wallerstein: Are you aware of the whole argument by Paul Tillich that guilt is a product of the sixteenth century and that previously there were other sentiments in Christianity, that that was a real shift that Protestantism brought in?[14] And what do you think of that in relation to what you've just said?

Hopkins: I think that it's not right that guilt is a creation of the sixteenth century. What is right is that, as everybody knows, Puritanism was one of the several reactions that occurred in Western culture against a previous libertinage.

Wallerstein: But it's not new. He makes a strong case that it is fundamentally different from the sentiments of earlier periods since it involved precisely incorporation into the superego. I don't think he used the word, 'superego', but anyway that was the sense.

Hopkins: Well, I don't know enough about sixteenth-century England, but what I have read reminds me very much of what [anchorites] say, I mean, the perpetual battle of will and desire.

Koenigsberger: I wouldn't believe Tillich for a moment ...

Burguière: It's over the same question. I want to know how Keith Hopkins can explain perhaps the combination of the permanency, the stability of this new moral paradigm with a valorisation of chastity, virginity, with the devalorisation of sexuality and the very fluctuant effect of the attitudes toward sexual norms. I would take two examples: [first] this strange question of birth-control practices. This, the practice of birth control has been forbidden by the Church. ... A new attitude, a new official phenomenon, by condemnation of the what they call the sin of Onan. And Philippe Ariès, for instance, has elaborated the hypothesis that this prohibition has been internalised like others, such as, for instance, incest or

[14] Paul Tillich, *The Courage to Be* (London: Nisbet, 1952).

something like that.[15] And that the techniques of birth control have always been only preserved in the illegal, abnormal practices, for instance, among prostitutes and so on. And that's this, the revival of the practices occurred with a change in morality, perhaps in Western Europe, as in France, for instance, during the eighteenth century. But now we have some qualitative indications that those practices existed for instance during the fifteenth and perhaps the sixteenth century. And for a generalisation, the statistical ratio of birth control we have now evidence for Geneva, at the end of the sixteenth century and perhaps for some parts of France even in the seventeenth century. So I have been pushing more of a fluctuant application of a prohibition and if we have no evidence for earlier periods, it's perhaps because we have to search. And because we miss statistical evidences. And if we look at another example, it's clear that between the sixteenth and seventeenth century all the attitudes of the whole population toward sexuality change, I would say, not only of the elite: first by the prohibition of public baths, prohibition of houses of prostitution, which were, for instance, during the fifteenth century, very officially managed by municipalities. [They] were closed at the turn of the sixteenth century. And with the beginning of statistical and demographical records we have very clear evidence that between the middle of the seventeenth century and the middle of the eighteenth century the rate of illegitimacy and the rate of prenuptial conceptions declined and nearly disappeared, and suddenly emerged at about the middle of the eighteenth century. And this gives an indication, I mean, of the attitude of the whole population which is perhaps somewhat different from that of the elites on which we have more qualitative evidence, for instance, on libertinage and so on. ... I have no answer to this question, but I think that there is really certainly the combination with an internal control by a new form of self-restraint created by a new configuration of power, and also an external control, the effect really of the control of the Church. You know, we have the impression that the Church elaborated a moral system of prohibition since the third or fourth century that was applied only perhaps ten centuries later if we [take the] prohibitions of marriage. All these prohibitions of marriage are really a very new phenomenon among Christianity. As radically new as the valorisation of chastity and so on. It seems that the real control of marriage prohibition because of consanguinity affinity and so on, was nearly impos-

[15] Philippe Ariès, 'Sur les origines de la contraception en France', *Population* 8: 3 (1953), pp. 465–72.—eds.

sible for the great number of the population till the end of the sixteenth century because of the lack of administration [of] all that. So that I think that in the fluctuation of the general moral attitudes, there is a mix of change in the outlook, in the morals, I would say, but also a changing of the general context of control, of public control. This is [some alternative for the] question of attitude towards sexuality.

McNeill: I have one hesitant suggestion. But it does occur to me. It seems to me the fluctuating patterns of population pressure on resources in the Roman Empire might have something to do—now I'm not prepared to make a clear proposition about this—might have something to do with the fluctuating attitudes toward sex and marriage. I don't know where the monks came from but I suspect, I had always supposed, I don't know, the Egyptian and East, at least the Roman monks came out of a very [...] landscape. And then you get the [Antonine] plagues,[16] very sharp cutback in population, the recovery, by the fourth century, of population and the generalisation of these prohibitions of sexual indulgence, coming to a sort of crystallisation in the fourth century, at a time when population was recovering, or had recovered from the loss of the Antonine plague. And if you try to think of peasants, which you really know nothing about it, that's the trouble: at least I know nothing about it. Perhaps you do. The problem of four or five children coming of age on a piece of land that cannot sustain the life of the peasant family—how do you marry all four of them without more land? What opportunities are there in town? What options are there? Well, going into the Church, going into this holy life of the monk is certainly one answer to the problem of not enough land to divide up amongst your sons. In later times—you probably know something about nineteenth-century Ireland, where this was carried to the most extraordinary extreme—postponing marriage, shipping out [emigrating], and going into the Church were the three things that you could do under these circumstances. Not at the time, it's after the plague, after the famine when the prudential control of the population became absolutely primary consideration for the Irish by this kind of adaptation. Now, it seems to me, in ancient Christianity in ancient Rome in the crystallisation of the Christian attitude toward sex and marriage, they may have been moving into—at least in the Eastern part of the Roman Empire, maybe not the West—a social situation like the Irish situation after 1845, where

[16]Antonine Plague, AD 165–180, suspected to have been either smallpox or measles; see W. H. McNeill, *Plagues and Peoples* (Garden City, NY: Anchor, 1976).—eds.

the population starts to creep up again after the catastrophe that had occurred 70–80 years before. And the unwillingness of the population, the peasant population, the lower classes, to face the catastrophe, the crisis that they, their forefathers have known, is resolved in this fashion. Now, this seems to me as too simple. But this is the playing it against the curves of the population, pressure on the land, what rural life was like, would be the place I would look for explanation. I don't know whether I could find it.

Hopkins: I think that's the difficulty. You couldn't. We don't have good population data over time for any part of the Roman Empire, either province or village. Well there was one. I think there's one village that's an exception, but it's terribly difficult to interpret. It does have the data straddling the Antonine plague. About 170, something like that. The difficulty is, it seems to me quite clear, that the Christians didn't fix their ethos in order to meet a demographic crisis. But, on the other hand, of course, at some stage, it seems to me likely that the population of the empire, say in AD 400, the population of the area which the empire covered in AD 700 was much less. In other words between 400 and 700 there was, I think, a significant decline in population. And the prohibitions against remarriage were probably more important than the significant aggregate numbers but small proportionate numbers who became celibates. So it's the prohibition of remarriage, rather than the encouragement never to marry, which I think would have had, over time, a significant effect. How much, say sheer poverty encourages people to undertake a life of complete poverty … one can't tell. What difference does it make, you know, to walk away from a small farm and stick yourself in a hut in the desert? God alone knows.

Wallerstein: I suggest: was it being pushed away rather than walking away.

Hopkins: Well, I mean, of course, there's the Christians; don't forget, the literature is hagiographic. So, they don't say 'Reluctantly I became a monk'!

Evers: If you want to play this game of taking the peasants and the peasant economy, I think one would, or ought to be able to, make just a counter argument: if constrained asceticism is on the upward, which leads into the institutionalisation in the forms of larger communities of monks, you need a peasant economy that can support the surplus population that is not productive …

Wallerstein: Not if they don't eat very much. [Laughter.]

Hopkins: … It depends. There's obviously a tension. The eremites themselves do seem to be supported by the local village. They are, in effect, non-begging beggars. They don't have to beg. People come and give them things. At least in the accounts that survive. The monasteries do make attempts to some extent, to be self-supporting and, of course, in the west, they become property owners and working monks. And, therefore in that sense, are really unmarried farmers. They may not be without children … But the difficulty is that obviously one can think up these things. What we can't do is solve them from the evidence available.

McNeill: No, I perfectly understand. This is the line which I would be trying to make a connection with a hypothesis.

Bogner: I don't know much about Roman history. Most of what I know about it is from the writings of Max Weber. But he wrote something which does seem to fit very well into what McNeill said and what you have just said—about the changing population in the late Roman Empire: Weber's argument was that the economy of the Roman Empire was centrally based on a slave economy and slavery is a kind of institutionalisation of labour force, which at least was not able to reproduce itself by means of biological reproduction, but had to be reproduced by wars in which slaves could be caught and brought into the markets of Rome. And he put forward—I don't know if it is from him—the hypothesis that increasing difficulties in the supplying of slaves to the slave markets brought about a situation in which it was necessary to change into another form of institutionalisation of labour force, a form which was able to reproduce itself by the biological way of reproduction, which meant members of the labour force had to be allowed a certain degree of autonomy, had to be allowed to marry, had to be allowed to have family life. And, in fact, this moral revolution, to me, seems to have provided the appropriate ideology for—if I may use this badly misused conception—for this transition in the institutionalised form of the labour force. And also the decrease in population and the decrease in labour force gives, according to Norbert Elias's theory of power balances, a relatively increased power to the labour force. You know, scarce labourers are better paid than abundant labourers and scarce slaves are better treated than abundant slaves. And this might have been the power basis of this transition.

Hopkins: Alas, it might have been but it's improbable that it was. Weber was writing, as Marx was, at a time when it was thought that the numbers of slaves in the Greco-Roman world were much larger than any scholar now legitimately thinks that they are. Slaves were important in

Roman Italy in, say, the last century BC. About 30 per cent of the population, according to one estimate, were slaves and, in other parts of the empire, in the central core, perhaps ten per cent. But it may have been more in Greece and in Western Asia Minor. But otherwise slaves, I think, were predominantly domestic ancillary workers for people who had been hit by death, rather than a major force in production. And the supply of slaves from foreign conquests can't have been great after the expansion of the empire stopped with the death of Augustus, apart from occasional wars in Britain and Romania and, say, against the Jews. So that I think that slaves, if they were agricultural slaves, were already settled with wives and reproducing themselves.

I think we've relied very much on the Jamaican evidence to see that slaves were non-biologically self-reproducing because they didn't then know the evidence from the American South or from Brazil when slaves were not merely self-reproducing but showed a dramatic rate of population growth. The evidence has been collected mostly since the Second World War. So I think that the general principle that slaves can't reproduce themselves is now shown not to be historically true. And we're left with either the Jamaican, where the ratio of males to females was sometimes as high as 17:1, as against the southern states of America, after the time that the import of slaves was prohibited to the time of the Emancipation, where the rate of population growth among slaves was enormous, and led to a migration of slaves from the eastern seaboard to the southwest.

Calvi: I'm going to move to a smaller argument I think. And I wanted to go back to that story you were mentioning, about the conversion to chastity of married couples, which I understand was a successful model pointing to a changing pattern in marital relationships. In the story, the initiative leading to mutual chastity was taken by women. Now this detail seems rather significant in that it stresses a meaningful role of women in marriage, which perhaps opposes the loss of juridical capacity you pointed to. And underlines their role as mediators in the spreading of the new faith. Now, Natalie Davis stressed the role of women in religious revivalism during the Reformation claiming … new forms of religiosity rather than new forms of autonomy and initiative, within specific social contexts.[17] Could chastity in the early centuries of Christianity provide new space for what has been called 'women's sphere', stressing values of

[17] Natalie Zemon Davis, *Society and Culture in Early Modern France: Eight Essays* (Stanford, CA: Stanford University Press, 1975).—eds.

autonomy, which were progressively undermined in secular society? In this sense, it would seem to me, that self-restriction would seem to acquire a more ambivalent meaning. Certainly not a merely negative one, if understood from the subjective standpoint of different social actors.

Hopkins: It's a lovely question. I agree completely with the first part and I should have mentioned Natalie Davis, as it were, as a parallel. Here, in my view ... the stories, I can't think of any exceptions, and they're *topoi*, they're highly conventionalised stories of seeking chastity. It isn't, in these marriage stories, it isn't the woman who universally seeks the initiative and, as I said, it seems to me the case that women can only maintain or heighten their status in the fourth century, I think, because of the greater incidence of physical violence and the lower frequency with which women can seek redress without male support in the courts, the only way in which women can maintain their status is by becoming chaste, either as permanent virgins or as abbesses, won by continence within marriage. The last part, the last sentence, as it were, was the twist in the tale. And how far from different actor's perception ... Certainly in the stories, the man sees the woman's initiative as highly threatening. One doesn't know whether the woman is afraid of sexual intercourse. I mean the famous essay by Freud about women's fear on the wedding night. And the story could at least have been motivated by that or as a desire to castrate the male. In other words, you're free to interpret these stories as you wish. What we don't know—because they're so heavily laden with hagiographic ideology, what we don't have any attempt, we can't see—is how the individuals actually interpreted this behaviour. We have this highly conventionalised account. What we don't have is any commentaries on those accounts!

Mennell: I think we probably ought to wind it up now. And I'd like to thank Professor Hopkins very much for producing such a first class discussion. It does seem to me that out of our session this morning has come a splendid illustration of closeness and interaction between micro- and macro-history, if you like. I think, perhaps, out of this, we shall want to ask a lot of questions in relation to Professor Elias's ideas. For example, the general ideas of the chains of interdependence lengthening and shortening and being patterned in complex ways and all the fluctuating balances of power that they involve, and the way in which these are related to changing patterns of constraint. We've mainly talked yesterday about the idea of increasing scale and increasing political centralisation. Here we've heard at least the possibility of a situation where the chains of interdependence were beginning to go into reverse. To what extent were the constraints on self-restraint reversed?

Civilisation, Culture and Power: Reflections on Norbert Elias's Genealogy of the West

Johann P. Arnason

1. It may be suggested that the notion of civilisation, as far as it appears in classical social theory, has a symptomatic rather than a systematic character: it emerges—primarily in the Durkheimian tradition—as a tentative response to problems and shortcomings of the dominant image of society. Following Touraine's analysis, we can summarise the main components of the latter as follows: the social world is subordinated to consensual normative principles, embodied in institutions; there is a more or less explicit tendency to portray society as a collective actor; this is facilitated by the tacit identification of society with the national state, while the specific problematic of states and inter-state relations is largely ignored; finally, the impact of normative societal structures on the individual is explained as assimilation through internalisation.[1] Although these ideas are by no means fully developed in Durkheim's work, their outlines are sufficiently

[1] Alain Touraine, *Le Retour de l'acteur* (Paris: Fayard, 1984). For Touraine, the Parsonian system is the most consistent and detailed version of the dominant image.

J. P. Arnason (✉)
Melbourne, VIC, Australia
e-mail: J.Arnason@latrobe.edu.au

A. Bogner, S. Mennell (eds.), *Civilisations, Civilising Processes and Modernity – A Debate*, Palgrave Studies on Norbert Elias,
https://doi.org/10.1007/978-3-030-80379-7_12

clear to make the supplements and potential correctives visible as well. Among the latter, the two-pronged guideline for a theory of civilisation stands out as particularly significant: it points on the one hand to contexts of interaction and communication beyond the horizon of a 'constituted society', to use Durkheim's phrase; on the other, it places the transformation of the human being in and by society in a broader perspective than the more restrictive model of integration through norms.

Lévi-Strauss's retreat from a theory of civilisation should be seen against this background. His answer to the problems of Durkheimian sociology involves a systematic 'decentring' of the concept of society; with the shift from social relations and historical transformations to more fundamental generative structures of the mind, the concept of civilisation also loses its former relevance. And despite the enormous gulf that separates both Marx and Weber from Lévi-Strauss, the vague and matter-of-fact usage of the notion of civilisation in one case and its fragmentation in the other can be related to the same issue: neither the Marxian nor the Weberian conceptual schemes are as easily assimilable to the traditional image of society as Durkheim's (notwithstanding their differences, an important feature of both is the inclusion of social relations in a broader anthropological context), and the need to compensate for its limitations is correspondingly less acute.[2] It seems to me that the same problematic—the tension between the concept of society and the concept of civilisation—is also the key to Norbert Elias's theoretical innovations. His analysis of the civilising process is not only an attempt to reactivate the classical project of a historical sociology, largely abandoned in the twentieth century; it is at the same time a strategy of reconceptualisation at the most basic level. As a result, he proposes an image of society that is most directly opposed to the mainstream paradigm of institutionalised and internalised norms, but can also be understood as an alternative to Marxist, structuralist and system-theoretical criticisms of the latter. Elias's idea of sociology is based on the concept of power, redefined and generalised so as to grasp the core of social life.

[2] Our analysis is confined to classical social theory and subsequent developments; Comte's idea of sociology and his notion of civilisation, much closer to eighteenth-century precursors, is beyond the scope of this paper. Elias's partial rehabilitation of Comtean ideas is only understandable in the light of his critique of Marx, Weber and Durkheim. [See Elias, chapter 1 in *What is Sociology?*—eds.].

This reorientation enables him to deal with the traditional problems of civilizational analysis from a new angle (as we shall see, Elias has developed an explanation of 'the rise of the West', as well as a less detailed but highly distinctive account of the Westernisation of the world); and conversely, the ability to generate a structural and dynamic theory of civilisation strengthens his image of society vis-à-vis others that have proved much less suitable for such purposes.

The search for networks of power and control as the common denominator of social structures and civilising processes is, however, not the first step in Elias's argument. In the first instance, he focuses on a more specific phenomenon: the historically changing standards of 'civilised behaviour'. From the viewpoint of the history of ideas, the most significant thing about this aspect of civilisation is that it has frequently been used to specify the contrast with culture, limited to the sphere of spiritual values, and with progress, defined in terms of technology and material life. Moreover, it has figured prominently in some attempts to link a national identity to a civilising mission. But Elias has a further and more important reason to choose this starting-point: the canons of civility can on the one hand be subsumed under the concept of norms as 'cultural definitions of desirable behaviour', yet on the other hand they are excluded or at least relegated to a marginal role by the prevalent conception of the social nexus as a moral structure. Elias's aim is to show that the civilizational periphery of normative structures is in the last instance reducible to configurations of power, and that this infrastructure also determines the overall framework of social life—including its normative patterns: the 'social constraint towards self-constraint' is seen as one aspect of the more comprehensive network of controls and constraints, arising from the interdependence of people. This is a 'deconstructive' operation *avant la lettre*: a dominant interpretation is undermined through a critique that derives its strength from a novel approach to previously neglected or marginalised phenomena.

Within the limits of this paper, it is obviously out of the question to discuss Elias's theory *in extenso*; I will confine myself to a few remarks on its basic presuppositions (sections 2, 3) and on some questions raised by the interpretation of European and post-European history (sections 4, 5).

2. One of the most conspicuous developments in contemporary social thought is the emergence of several concepts and theories of power that address themselves more or less directly to a common task: a 'war on two fronts' against Durkheimian (or 'Durksonian', to use an American *bon mot*) and Marxian images of society—that is, against the conception of a

social order based on norms and values, as well as against the one-sided emphasis on economic determinants and their internal dynamic. This trend is as diversified as it is widespread. There is, for example, an obvious and fundamental difference between those who analyse power in terms of resources, distribution, organisation and productivity, thus using a conceptual apparatus that is at least partly modelled on the Marxian interpretation of economic structures, and other theorists who are more interested in the cultural definitions and imaginary dimensions of power. Foucault and Giddens are the most prominent representatives of the first approach. The second is exemplified by Lefort's renewal of political philosophy, centring on the political constitution of society in the triple sense of a *mise en forme*, *mise en sens* and *mise en scène* of human coexistence;[3] the main themes of Touraine's work—the capacity of society to create its own rules and institutions, the cultural model as a regulative image of this capacity, and the struggle between social movements that strive to control it—have similar connotations. Bourdieu's project can perhaps be understood as an attempt to synthesise the two strands, but it is heavily weighted in favour of the first: a more systematic generalisation of Marxian categories—primarily capital and class—beyond their original context is expected to pave the way for an analysis of the cultural field as a symbolic transfiguration of power relations.

In retrospect, Elias's analysis of the civilising process appears as a pioneering contribution to a general theory of power. Regardless of the differences between his views and those of the above-mentioned theorists (some instructive contrasts will be noted in the following discussion), some common presuppositions are more easily recognisable in the original model than in later variants. In particular, three preliminary steps are clearly outlined: the generalisation, pluralisation and historicisation of the concept of power.[4]

[3] Claude Lefort, 'Permanence du Theologico-politique?', *Le temps de la reflexion*, II (Paris, 1981), pp. 13–60.

[4] The interpretation proposed here draws mainly on the following works by Norbert Elias: *The Civilizing Process*, vol. 1, *The History of Manners* (New York: Urizen, 1978); vol. 2, *State Formation and Civilization* (Oxford: Blackwell, 1982) [*On the Process of Civilisation* (Dublin: UCD Press, 2012—Collected Works, vol. 3)]; *Die höfische Gesellschaft* (Neuwied: Luchterhand, 1970) [*The Court Society* (Dublin: UCD Press, 2006—Collected Works, vol.2)]; *Was ist Soziologie?* (München: Juventa, 1970) [*What is Sociology?* (Dublin: UCD Press, 2012—Collected Works, vol. 5)]; 'Sociology of Knowledge: New Perspectives', *Sociology* (1971), pp. 149–68 and 355–70 [in *Essays I: On the Sociology of Knowledge and the*

(a) For Elias, the concept of power can denote a relationship between men and natural objects, as well as between men; by implication, it is also applicable to the relationship between social individuals and their own malleable, but neither amorphous nor self-regulating nature. The manifestations of power in this general sense are synonymous with the exercise of control. Elias refers to a 'triad of basic controls': a society can be analysed and its level of development defined in terms of its capacity to control natural processes and social relations, and to make its individual members capable of controlling their own behaviour.[5] Although he characterises the basic controls as universals of social life, he does not try to construct—à la Parsons—a closed scheme of universal functions; as his later distinction between means of production, means of violence and means of orientation shows, the model is still open to some revisions and alternative formulations. But the emphasis on orientation as a specific activity, irreducible to production, does not *eo ipso* entail a more limited use of the concept of power. Elias speaks of a basic 'need for control and orientation', and he describes cognitive progress as a 'changing balance of power' between subject and object.[6] Positions and constellations of power thus remain the most global frame of reference for any classification of human ways of relating to the world. The notion of orientation as a modality of power sets Elias's sociology of knowledge apart from some later interpretations; he does not need bridging concepts—such as 'cultural capital'—to establish a link between power and culture, but he also tries to avoid the short-cut suggested by those who equate the cognitive expressions of power with instrumental reason.

Sciences (Dublin: UCD Press, 2009—Collected Works, vol. 14), pp. 1–41]; 'Zur Grundlegung einer Theorie sozialer Prozesse', *Zeitschrift für Soziologie* 6: 2 (1977), pp. 127–49 [in *Essays III: On Sociology and the Humanities* (Dublin: UCD Press, 2009—Collected Works, vol. 16), pp. 9–39]; 'Über die Zeit', *Merkur* 411 (1982), pp. 841–56 and 412 (1982), pp. 998–1016 [*An Essay on Time* (Dublin: UCD Press, 2007—vol. 9)]; 'Der Rückzug der Soziologen auf die Gegenwart', *Kölner Zeitschrift für Soziologie and Sozialpsychologie* 35 (1983), pp. 29–40 ['The retreat of sociologists into the present', in *Essays III*, pp. 107–26]; *The Established and the Outsiders* (London: Frank Cass, 1965), with J. L. Scotson [enlarged edn., Dublin, UCD Press, 2008—Collected Works, vol. 4]. See also Johan Goudblom, *Sociology in the Balance* (Oxford; Blackwell, 1977); and Artur Bogner, *Macht und Herrschaft unter zivilisationstheoretischer Perspektive* (unpublished ms, Bielefeld, 1981).

[5] Elías, *Was ist Sociologie?*, p. 173 [*What is Sociology?*, pp. 151–2].

[6] Elias, 'Sociology of Knowledge: New Perspectives', p. 23.

(b) Although Elias has not worked out a systematic typology of the sources of power, his remarks on their 'polymorphous' character are specific enough to set the course for further exploration. To begin with the very diversity of the 'basic controls' involves the foundations as well as the uses of power: to the extent that the various mechanisms of control are monopolised by specialists and/or privileged minorities, they become the bases of economic, military, political and cultural power structures. The distribution of power is further affected by the invention or discovery of new means of control. Over and above the recurrent functions of social life, all factors that can give rise to relations of interdependence—objects and skills, beliefs and traditions, needs and emotions—can also impart to them a more or less pronounced asymmetry and thus act as sources of power. This structural heterogeneity has another crucial implication: the normal pattern of social relations is not the unconditional and exclusive possession of power by rulers, elites, or dominant classes; it is rather a plurality of more or less fluctuating balances of power, where the multiplicity of resources and mutual needs enables the weaker sides to establish minor but by no means irrelevant positions of power. Elias thus drew attention to the phenomenon that Anthony Giddens later described as a 'dialectic of control'.

(c) This comprehensive and at the same time highly differentiated conception of power guides the analysis of its historical transformations. Elias's main concern is with two interrelated long-term processes, the formation of states (on the basis of the twin monopolies of violence and taxation) and the increasingly strict rules of self-restraint which a more complex and closely knit society imposes on its members. But from a more general point of view, his theory allows for changes in the relationship between basic functions and types of power (military, political, economic etc.), transformations brought about by the accumulation and diversification of resources, as well as increases or decreases in the power differentials between interdependent groups (the modernisation of European societies has, as he sees it, led to a triple reduction of such differentials: between rulers and ruled, between higher and lower strata, and as a result of the general trend towards more reciprocal and multipolar forms of control).

Elias's account of the metamorphoses of power touches on several points that have been neglected or only partially thematised by later theorists. As he shows, the long-drawn-out development of a monopoly of physical violence and the concomitant elimination of violence from wide areas of social life are the most fundamental preconditions of later changes; this insight has certainly not been fully assimilated by contemporary theories of power. Another key component of his thesis is the generalised model of competition and monopolisation, applicable to pre-capitalist societies and non-market spheres of action no less than to a capitalist economy. The diagnosis of a long-term trend towards the centralisation of power and the amalgamation of various monopolies is especially significant when seen in conjunction with another phenomenon: the increasing relative weight of constraints inherent in the mechanisms of interdependence, as against the power positions of concrete groups and individuals. Elias does not deal with this issue at length, but he is clearly aware of its significance.[7] Interconnections, tensions and alternations between the concentration and the depersonalisation of power have certainly played an important role in modern and contemporary history (it might perhaps be suggested that an analysis of the changing patterns of Soviet power—from the dictatorship of Stalin, and from the post-Stalin crisis to the reconsolidation of the apparatus—would be particularly revealing in this respect).

Finally, Elias's analysis of the civilising process has some bearing on contemporary discussions about the origins and distinctive characteristics of capitalism. Anthony Giddens has reinterpreted the difference between capitalist and pre-capitalist forms of society, emphasised but never fully clarified by Marx, in the perspective of a general theory of power: pre-capitalist societies are characterised by the predominance of authoritative resources—that is, those involved in the production and maintenance of societal power, whereas the capitalist epoch shifts the balance towards the primacy of allocative resources, the social appropriation of nature; on this view, Marx's paradigm of production represents an unjustified extension of the latter category. The comparable interpretation that can be extracted from Elias's writings qualifies the contrast in several respects: the industrial revolution and the expanding market economy are defined as aspects of a more general process of accelerated differentiation, while corresponding processes of integration lag far behind. The preconditions for this unbalanced change were, however, created by a prolonged process of

[7] Elias, *Was ist Soziologie?*, p. 98 [*What is Sociology?*, p. 89].

integration whose most massive result was the formation of national states in Western Europe; inasmuch as the strengthening of the industrial state and the worldwide diffusion of its institutions and techniques have been characteristic of recent history, it can be argued that the temporary advance of differentiation is being offset by new forms of integration.

3. On the basis of the above summary, we can spell out some distinctive features of the critique which Elias directs against Marx and Durkheim. In both cases, his strategy involves—at least implicitly—an attempt to appropriate genuine insights and distinguish them from misinterpretations. Marx's model of the social world as a hierarchy of fundamental and derivative structures is accepted and invoked against his overvaluation of the economic sphere. According to Elias, the emergence of the modern capitalist economy presupposes a more fundamental process: the overall pacification and rationalisation of social life, Durkheim's emphasis on social constraint is retained and even accentuated, but not so his tendency to identify it with normative regulation; the quintessence of Elias's argument is his claim to have identified a deeper level of the interplay between social and individual phenomena. The development of internalised controls in response to the pressures and needs of social interdependence does not presuppose a normative framework; rather, a certain extension of 'pacified social space' and a corresponding degree of the 'muting of drives' is a necessary prerequisite for the imposition of norms on social figurations.

However, my main purpose in recapitulating Elias's theory of power is not to explore contrasts and parallels with other interpretations. In the present context, it would seem more rewarding to examine some loose ends and open questions that are particularly relevant to the problematic of civilisation. I shall, in other words, try to show that there is some scope for arguing 'with Elias against Elias'; if the range of his analysis throws some implications of a general theory of power more sharply into relief than later versions mostly do, this very thoroughness also makes it easier to determine the intrinsic limits of the theory and detect the potential grounding for a broader perspective.

To begin with, let us take another look at the concept of power. Elias's understanding of it as referring to a permanent and omnipresent aspect of the human condition differs in one very decisive respect from other similar arguments: power in the most general sense is not equated with the transformative capacity inherent in human action or praxis (this is, e.g. the line taken by Anthony Giddens in his more recent writings).[8] Elias rejects the 'action frame of reference' as dependent on the notion of *homo clausus*—that is, a human nature, a human essence, or an aggregate of human competences isolable from the relations between human beings (and, although this aspect is less developed in Elias's work, relations between human beings and the world). His redefinition of power is—*mutatis mutandis*—based on the same considerations as his critique of the traditional notion of development: it should be transformed from an action concept into a functional one—with the obvious proviso that the concept of function also needs to be revised and linked to a multipolar field of relations, rather than to a unified and self-maintaining system.[9]

Elias proposes a paradigm shift from *homo clausus* to *homines aperti*—that is people situated in contexts, subject to constraints, and seeking to maximise control.[10] From this point of view, the phenomenon of power is marked by a fundamental ambivalence: it involves the experience of constraints and the need for control. The interconnection is particularly obvious in the field of social relations, where the control of one side is constraint for others. But as we have seen, Elias also interprets the encounter between humans and nature in terms of power relations; and if the growth of knowledge shifts the 'balance of power' towards the object, the improvement of the means of orientation can also be seen as a reinforcement of cognitive constraints. As for the control over internal nature, Elias has demonstrated at great length—and very convincingly—that it is achieved through the internalisation of constraints.

This approach to power suggests a further conclusion that would already take us beyond Elias's explicit theses; the generalisation of the concept paves the way for its relativisation. The ambivalence of power lays it open to interpretations. To determine the meaning of constraints and controls as well as the relationships between them, varying across the dimensions of the human context—nature, society, and the unconscious, to mention only the major spheres—and changing in the course of history, is the task of cultural models and one of the focal points of their divergences. If we follow Hans Blumenberg's analysis of mythic structures, the

[8] See for example Anthony Giddens, *The Constitution of Society* (Cambridge: Polity, 1984).—eds.

[9] Elias, *Was ist Soziologie?*, p. 162 [*What is Sociology?*, p. 142].

[10] Elias's most extensive discussion of this is in the 1968 Postscript to *Process of Civilisation*, pp. 512–26.—eds.

primordial patterns of interpretation are shaped by the simultaneous impact of the 'absolutism of reality' and the 'absolutism of wishes'—that is, the unmediated encounter between equally absolute projects of constraints and control.[11] The later development of world-views leads to differentiations on both sides as well as in their interrelations, but not to the drawing of definitive boundaries. Models of control are transferred and extrapolated from one domain to others; from this point of view, traditional projections of the social order onto nature and modern visions of a social life subjected to the same kind of mastery as natural processes are comparable. Rejections or problematisations of established patterns of control in one area can also stimulate parallel developments in others, without necessarily involving a structural assimilation. Thus, the reaction against technocratic models of societal rationalisation has raised questions about the underlying paradigm of rationality and its adequacy vis-à-vis nature as well; the changes in conceptions of individual autonomy and self-control that have resulted from the discovery of the unconscious also affect the perspectives of human emancipation in regard to the natural and societal context.

In short, the multiform and changing relationships between constraints and control must be examined through the prism of cultural interpretations that alternate between the sharpening of distinctions and the codification of more or less comprehensive models. In the last instance, the general use of the term 'control' becomes questionable: What is at stake is rather the regulative idea of autonomy, incorporating the varying kinds and degrees of mastery that correspond to different aspects of the human context.

Elias's discussion of the basic concepts of sociology contains a further hint that seems relevant to this issue. One of his main objections to the conventional concept of function is that it obscures 'the multi-perspectival character' of social relations.[12] His own redefinition is designed to do justice to this fundamental trait of social reality. But to talk about perspectives is to talk about interpretations. Moreover, the 'multi-perspectival' pattern should apply to the basic functions no less than others should. Elias's references to 'society' as the bearer are difficult to reconcile with his statements about the irreducible plurality and multipolarity of human figurations.

[11] Hans Blumenberg, *Arbeit am Mythos* (Frankfurt: Suhrkamp, 1979).

[12] Elias, *Was ist Soziologie?*, p. 137 [*What is Sociology?*, p. 122].

4. The line of argument that I have sketched would lead to a more concrete analysis of the interpretations, cultural definitions, images and even phantasms of power. This problematic plays only a very marginal role in Elias's work. Suggestive hints are not hard to find, but they do not add up to a coherent model.[13] For our present purposes, an enumeration of such pointers would not be the most useful strategy; instead, I shall try to single out some nodal points in the analysis of the civilising process and show how they indirectly illustrate the need for a closer look at cultural contexts, both as direct determinants and as the background of other factors.

As noted above, Elias's theory of civilisation is—among other things—an attempt to explain the rise of the West and its impact on the non-Western world. If we want to reduce his thesis to its barest outlines, it places the enforced, in its entirety uncontrolled, but nevertheless increasingly rational pursuit of power at the centre of both European history and its worldwide sequel. In more specific terms, the relationship between societal power, which is ultimately based on the monopoly of violence, and the conquest of nature which is progressively achieved through the differentiation and technicisation of labour, has undergone a radical and unique change in the West. Elias refers to the 'transformation of the West into a society where more and more people earn a living through occupational work'.[14] This process has two sides: on the one hand the gradual integration of the dominant social groups into the division of labour, the

[13] The concluding remarks of the essay on time (*Über die Zeit*) are particularly interesting. After reducing the experience and interpretation of time to a practical foundation, the measurement and coordination of concrete processes, Elias tries to explain why this primary meaning of temporality is overlaid and distorted by the 'substantialising' interpretations that have [marked] philosophical theories of time. His answer is that the idea of time *per se* as an unchanging and universal structure, encompassing all specific processes, can only be understood as the expression of a deep-seated and tenacious human need: the quest for a permanent reality as a counter-weight to human finitude and transitoriness. The conception of time as the unchanging background of change is thus ultimately rooted in the same way of thinking as theological notions of eternity. [See Elias, *An Essay on Time*; although he wrote the book in English, it was first published in Dutch translation (1974–5), then in German in 1984, about the time of the conference, and the first English edition appeared in 1992.—eds.].

[14] Elias, *Process of Civilisation*, p. 260. In the original German version (*Über den Prozess der Zivilisation*, Bd.2, Frankfurt: Suhrkamp, 1977, p. 72), a stronger formulation is used: 'Umbildung des Abendlandes zu einer als Ganzes arbeitenden Gesellschaft'.

emergence of 'more constrained upper classes',[15] on the other the 'slow rise of lower, working, urban strata to political autonomy and finally—first in the form of the professional middle classes—to political leadership'; the latter development 'provides the key to almost all the structural peculiarities distinguishing Western societies from those of the Orient, and giving them their specific stamp'.[16] The Western pattern of technological progress and 'the evolution of money to that specific form of "capital" which is characteristic of the West'[17] are further aspects of the same overall trend. Although the above quotation refers to the Orient, Elias is clearly more interested in the contrast with the classical world, based—as he sees it—on slavery; he identifies the beginnings of a distinctively Western pattern with the more dynamic phase of the Middle Ages.[18]

A configuration of power and its built-in developmental trend are thus the main themes of Elias's analysis. This choice is reflected on the level of basic concepts and methodological principles. To situate Elias's approach in the context of current debates, the distinction between organic constraints (including ecological, technological and economic structures) and logical constraints (including general rules of coherence as well as more specific patterns of meaningfulness)[19] seems useful. In contrast to this dichotomy, Elias focuses on figurational constraints—that is, those inherent in the interactions and interdependences of social individuals. And this refusal is combined with a rejection of systems theory, whose basic flaw is—according to Elias—the surreptitious reintroduction of a teleological perspective; no model of purposive rationality or consensual integration can do justice to the unplanned 'order *sui generis*' which arises from the 'immanent regularities of social figurations'. The proposed alternative to both action theory and systems theory is best exemplified by Elias's use of the concepts of differentiation and integration; he avoids the teleological

[15] Ibid., p. 466.

[16] Ibid., p. 253.

[17] Ibid., p. 261.

[18] As far as we can judge from Elias's brief excursus on the classical world (ibid., pp. 258–63), he seems both to overestimate the role of slavery (this argument generalises some features of the late Roman republic) and to rely on questionable analogies with a capitalist economy to explain its consequences.

[19] For a succinct formulation of this distinction, see Aram A. Yengoyan, 'Cultural forms and a theory of constraints', in A.L. Becker and A.A. Yengoyan, *The Imagination of Reality: Essays in Southeast Asian Coherence Systems* (Norwood, NJ: ABLEX, 1979), pp. 325–30; this article draws on the work of Clifford Geertz.

connotations which they have acquired in Parsonian systems theory by linking them closely with processes of competition and monopolisation. The latter are not interpreted as modes of strategic action, but rather used to illustrate the trans-intentional dynamics of interaction. It should be noted that the two conceptual pairs are not parallel: the development of integrative functions is no less affected by competition than is the corresponding process of differentiation.

For Elias, however, the figurational constraints are not simply an additional object of sociological inquiry. His analysis of them is at the same time a sustained attempt to absorb or subordinate the other types of constraints. This line of argument entails, as I will try to show, some highly selective and questionable interpretations of the historical evidence.

The one-sided accentuation of certain determinants and the systematic minimisation of others is not only apparent in Elias's reconstruction of the civilising process; it also influences his account of its initial conditions. While the role of the Greco-Roman heritage as a general background resource is not disputed, Elias insists that the ways of using it and the problems to which it is applied are specific to the new society that takes shape in the Dark Ages and enters a phase of accelerated development in the eleventh century. On this view, there are no fundamental differences between earlier and later revivals of the classical legacy; all 'renaissances' are equally determined by their real historical environment.

This conception of the relationship between the ancient and the mediaeval world excludes at least two major issues from consideration. On the one hand, it can be argued that the latent presence of the Greco-Roman heritage—simplified, largely forgotten for a long time and reactivated only under the influence of a more advanced civilisation (Islam), but never completely erased—modified the interrelations of culture and society in a way hardly paralleled elsewhere: the scope and variety of the cultural potential, at first incommensurable with a drastically impoverished and de-differentiated society, is in the long run—and given favourable conditions—conducive to exceptionally vigorous outbursts of social creativity. On the other hand, there are some reasons to believe that a mutually destructive interaction between the decadent Roman Empire and the barbarian societies led to what Franz Borkenau has described as a 'self-propelling process of moral disintegration'.[20] The result seems to have

[20] Franz Borkenau, *End and Beginning: On the Generations of Cultures and the Origins of the West* (New York: Columbia University Press, 1981), p. 385.

been a general regression of social life; the specific barbarism of the Dark Ages is the outcome of this process, not simply the normal condition of a society which lacks both a complex division of labour and a centralised monopoly of violence.

The shift from a coastal to a continental basis was a major achievement of early mediaeval society; the evolution of inland forms of communication was, as Elias points out, a crucially important prelude to later stages of European expansion. But he sees what Lynn White calls the 'northward shift of Europe's focus'[21] primarily as the result of social processes; the specific ecological constraints and possibilities brought into play by this broadening of the geographical basis are not examined. Nor is the many-sided technological progress which accompanied the internal expansion—the introduction of the heavy plough, the more extensive and diversified use of water power, and the development of mining, to mention only the most important aspects—considered as a formative factor.

It might not be very difficult to supplement Elias's analysis of mediaeval social processes with a more detailed account of ecological and technological contexts. The economic structures pose a more serious problem. According to Elias, the development and expansion of the monetary nexus is an essential component of the civilising process, but he rejects the absolute distinction between a natural and a monetary economy and insists that the gradual monetisation of economic life should be seen as only one aspect of the overall trend towards a more differentiated and integrated society. What thus disappears from sight is the possibility that the involvement in a monetary economy that had its centre outside the West might at one stage have been decisive for the development of the Western world. This question arises primarily with regard to the interconnections between Islam and mediaeval Europe. Maurice Lombard has—to my mind very convincingly—stood the Pirenne thesis on its head (in contrast, Elias seems to take it for granted) and shown how the Islamic conquests gave rise to an international system *sui generis* that had a deep and long-lasting impact on the less developed societies of Western Europe. This was not so

[21] The societal consequences of the 'technological boom', as Gimpel calls it, between the tenth and the thirteenth century are certainly not easy to trace, but various historians have put forward suggestive hypotheses; see particularly Lynn White, ***Medieval Technology and Social Change*** (Oxford: Oxford University Press, 1962); Jean Gimpel, ***The Medieval Machine: The Industrial Revolution of the Middle Ages*** (New York, Holt, Rinehart & Winston, 1976); William McNeill, ***Venice: The Hinge of Europe, 1081–1797*** (Chicago: Chicago University Press, 1974), and ***The Shape of European History*** (Oxford: Oxford University Press, 1974).

much a system of states as a system of markets, monetary movements and urban communities.[22]

But as noted above, Elias's neutralisation of the cultural field is the most important and also the most problematic part of his attempt to generalise the theory of power. The result of this operation is a view of European history that seems to leave no place for a relatively autonomous, let alone a 'pace-setting' development of world-views. Against all theories that focus on the role of interpretive frameworks or intellectual resources—whether they stress the direct and indirect implications of Christianity, the expansion of knowledge that began in the Middle Ages and gathered momentum in later centuries, or a supposedly universal trend towards a rationalisation of the world-view, accelerated by some specifically Western factors—Elias argues for an explanation in terms of the infrastructures and metamorphoses of power.

The marginalisation of culture was, as we have seen, foreshadowed in Elias's paradigm of the human condition. The subsumption of all fundamental relations under the imperatives of control excludes in principle the very idea of a radical autonomy of culture. If the cultural dimension is to be incorporated into the projected general theory of social processes, it can only figure as a peripheral offshoot. But on the other hand, some specific characteristics are required to distinguish it from other spheres. Because the a priori assimilation of interpretive and cognitive structures to the general dynamics of power dilutes all distinctive features, Elias has to introduce a compensatory principle: as he sees it, the analysis of social consciousness—that is, the sociology of knowledge in the broadest sense—should above all be concerned with cognitive progress as a collective enterprise. The primacy of process is thus asserted on the cultural level; but since the growth of knowledge is interpreted as a progressive approximation of cognitive symbols to objective reality the emphasis is clearly—in contrast to Elias's uncompromisingly anti-teleological view of social processes—on a teleological direction, rather than on internal mechanisms of the process. More precisely: the former compensates for the indeterminateness of the latter. Clearly, Elias does not want to commit himself to

[22] See Maurice Lombard, *The Golden Age of Islam* (Amsterdam: North Holland Publishing, 1975), and *Espaces et réseaux du haut moyen âge* (Paris: Mouton, 1972). Lombard's position is best summed up in the following statement from the latter book (p. 44): 'Ainsi nait et se développe dans l'Occident barbare une tendance qu'il *ne portait pas en lui-même*: éveil commercial, gonflement démographique, développement urbain, économie en progrès. L'origine doit en être cherchée dans le monde musulman.' (Italics mine, JPA).

the correspondence theory of truth—he dismisses the philosophical concept of truth as an 'anachronism'—but when he uses such expressions as 'nearness to reality' (*Realitätsnähe*) and 'objective adequacy' (*Objektadäquanz*), it is hard to see how he can avoid its well-known problems.[23]

The reintroduction of a teleological model and the reluctance to face its philosophical implications explain Elias's preference for concrete examples that would at least establish a prima facie case for his approach. In particular, one episode in the history of science is singled out as an unambiguous illustration of the general developmental trend, a thoroughly tested premise of further scientific progress and—last but not least—a model for the still immature science of society. This is the shift from the geocentric to the heliocentric world-view. Its significance for Elias is best summed up in the following statement:

> The development of the idea that the earth circles round the sun in a purely mechanical way in accordance with natural laws—that is, in a way not in the least determined by any purpose relating to humankind, and therefore no longer possessing any great emotional significance for people—presupposed and demanded at the same time a development in human beings themselves toward increased emotional control, a greater restraint of their spontaneous feeling that everything they experience and everything that concerns them takes its stamp from them, is the expression of an intention, a destiny, a purpose relating to themselves. Now, in the age that we call 'modern', people have reached a stage of self-detachment that enables them to conceive of natural processes as an autonomous sphere operating in a purely mechanical or causal way without intention or purpose or destiny, and having a meaning or purpose for themselves only if they are in a position, through objective knowledge, to control it and thereby give it a meaning and a purpose'.[24]

In this perspective, the Copernican revolution is the decisive step towards a rationalisation and modernisation of interpretive patterns, and at the same time the most convincing proof of the dependence of intellectual development on the underlying interplay of social constraint and self-restraint: the mechanisation of the world-view was prepared and prompted

[23] Elias, *Was ist Soziologie?*, p. 21 [*What is Sociology?*, pp. 18–19.] In English, Elias uses various terms in various places, but as often as not 'reality congruence' which, in *Involvement and Detachment* (Dublin: UCD Press, 2007 [Collected Works, vol. 8]), p. 135, he defines in terms of its 'cognitive value'.—eds.

[24] Elias, *Process of Civilisation*, p. 521.

by a long-drawn-out maturation of the external and internal mechanisms of control. Other modern innovations—such as the aesthetic experience of nature and the intensified self-observation of the individual—are, as Elias sees it, rooted in the same socio-psychological context. But the internalisation of controls also gives rise to the image of the closed and self-sufficient individual that has been such an enduring obstacle to the scientific study of society. This might perhaps be seen as Elias's version of the 'paradox of civilisation'.

Elias is undoubtedly right in stressing the central role of the Copernican revolution. This specific and at the same time sweeping change in the Western world-view—it represents a particularly close encounter of a concrete explanatory model with a general attitude to nature on the one hand, the self-understanding of humanity on the other—has not been given its due by evolutionary theories that tend to locate major changes at a much more abstract level. But if the exceptional significance of the phenomenon is beyond doubt, a closer look at its historical antecedents and consequences casts doubt on Elias's interpretation. The complexity and ambivalence of the Copernican revolution are convincingly demonstrated in Hans Blumenberg's recent analysis.[25] Far from being a simple and unequivocal transition from a subject-centred to an object-centred perspective, Copernicus's vindication of the heliocentric model must be seen against the background of the cultural significance which Greek and Christian tradition—partly synthesised, partly conflicting—had conferred upon astronomic issues; it disturbs a traditional conception of 'anthropological–cosmological interdependence', but does not *eo ipso* and at once replace it with a new overall interpretation; the revision of the Ptolemaic model can be understood either as an affirmation of human autonomy—in the sense of an ability to acquire valid knowledge through the use of reason—or a destruction of illusions about the place of human beings in the cosmos. This anthropological ambivalence is further confirmed by the various later theories that have used the Copernican analogy to demarcate or legitimise themselves; they range from Kant to Freud.

[25] Hans Blumenberg, *Die Genesis der kopernikanischen Welt* (Frankfurt: Suhrkamp, 1975); see p. 416: 'in the end, Copernicus triumphed not so much over his opponents as through them'. In other words, the most important aspect of the Copernican revolution was the conflict of interpretations which it generated. [English translation by Robert M. Wallace, *The Genesis of the Copernican World* (Cambridge, MA: MIT Press, 1987)—eds].

The claim that Elias's perspective on the growth of knowledge is more teleological than his model of the civilising process should not be misunderstood. Obviously, I am neither suggesting that he sees cognitive progress as the implementation of a conscious design nor am I arguing that the very idea of a directional process implies a teleological point of view; Elias's refutation of the later thesis must be regarded as conclusive. Between the narrow definition of teleology as the realisation of an intention and the extension of the concept to cover all long-term tendencies, however, there is an intermediary notion that is clearly relevant to Elias's reformulation of the agenda for a sociology of knowledge: the gradual approximation to an ideal final state. If the adequate representation of reality is not simply a normative criterion but the built-in goal of a collectively organised activity, the historical development of the latter must follow a more clear-cut and predetermined pattern than societal transformations. It might be objected that Elias's analysis of the civilising process also culminates in the vision of a future state of perfection: a lasting balance between the demands of social life on the one hand, personal needs and inclinations on the other. But this utopian extrapolation has no interpretive or explanatory functions; the dynamics of state formation and the corresponding modifications of personality structures are not guided by a striving for overall equilibrium. The perfect civilisation is a possible outcome of the process, rather than a regulative principle.

In the 1971 essay on the sociology of knowledge, Elias did not stress the role of symbolic structures as the medium of knowledge.[26] While his recent emphasis on this point, coupled with a clearer distinction between the means of orientation and the means of technological and political control, undoubtedly represents a significant innovation, it does not by itself entail a break with the teleological conception of knowledge. If cognitive processes are mediated by symbols, they can no longer be interpreted as direct contacts between consciousness and reality or as steps towards the elimination of all barriers between subject and object; but as long as they are analysed in terms of an overall trend towards 'object-adequacy' (even if the latter is now envisaged as a state of congruence rather than

[26] Elias seems to have recognised this gap: *The Symbol Theory* was published in journal articles in 1989, and posthumously as a book in 1991, and finally, including some passages on which he was working in the days before he died, as volume 13 of the Collected Works on 2011.—eds.

correspondence, so as to avoid the undesirable philosophical connotations of the latter term) the notion of an immanent goal is retained.

There is, however, another aspect of Elias's theory that might help to solve the problem. The proposed paradigm shift from *homo clausus* to *homines aperti* would result in a far-reaching reorientation of the social sciences (one way of exploring its consequences would be to compare it with Habermas's critique of the restrictive models and assumptions derived from the philosophy of consciousness or with Luhmann's attempt to grasp the specific openness of socio-cultural systems); Elias's concrete analyses have by no means exhausted its potential, and it could even be argued that he has in some respect toned down its implications. His rejection of Kant and the Kantian tradition in the theory of knowledge is clearly related to the more general critique of *homo clausus* (the transcendental subject is an avatar of the latter), but it stops short of the more strictly philosophical questions that must be raised if this critique is to achieve its aim. In the first place, the demolition of *homo clausus* would be incomplete without a rehabilitation of experience as an open-ended and creative confrontation between the human being and world. If Kant's theory of knowledge can be regarded as a particularly forceful statement of some ideas inherent in the closed image of man, the counter-image Elias wants to develop has obvious affinities with the philosophical currents that have challenged transcendentalism on its own ground; Adorno's emphasis on experience rather than form as the key to subjectivity is no less relevant than Gadamer's hermeneutical radicalisation of the concept of experience.[27] But the other side of what Gadamer calls the essential openness of experience is the autonomy of creative interpretation. If the philosophical premises of the sociology of knowledge—or, in more general terms, the sociology of culture—are redefined along these lines, we must recognise the fundamental interconnectedness of experience and interpretation as well as their mutual irreducibility and the utopian, but ineradicable aspiration to a complete fusion. Moreover, the very notion of interpretation is, as the development of hermeneutical philosophy has shown, inseparable from that of a conflict of interpretations. If all this is taken into account, the outlines of a plausible

[27] Theodor Adorno, *Gesammelte Schriften*, vol. 10/2 (Frankfurt: Suhrkamp, 1977), p. 752: 'The key contribution of the subject to knowledge is experience, not form'; and H.-G. Gadamer, *Truth and Method* (London: Sheed & Ward, 1975), p. 319: 'The dialectic of experience has its own fulfilment not in definitive knowledge, but in that openness to experience that is encouraged by experience itself'.

non-teleological paradigm begin to emerge: a civilizational approach to the sociology of knowledge would focus on the complex interrelations and long-term transformations of experiential, interpretive and dialogical capacities. Further discussion of this perspective—and the corollary task of reformulating the notion of cognitive progress—is beyond the scope of the present paper. But it is clearly as incompatible with Elias's wholesale dismissal of philosophy as with the reduction of the human condition to a network of controls and constraints.[28] I have raised some objections to Elias's excessive claims on behalf of the theory of figurations (more precisely: the configurations of power); to round off the discussion, it remains to show how the underestimation of other aspects—primarily the cultural context—affects his own thesis. I shall, in order words, try to pinpoint some weak links in his argument that illustrate the need for a more balanced approach.

First, however, some further implications of Elias's analysis should be noted. Among the most striking is the strong emphasis on the continuity of European history: 'One can readily imagine that, viewing the development of this whole period of human society, the medieval and modern periods together, later ages will see them as a single unified epoch, a great "Middle Age"'.[29] Compared to Jacques Le Goff's later concept of the 'long Middle Ages', lasting from the crisis of the Roman Empire in the third century to the democratic and industrial revolutions at the end of the

[28] The shift from *homo clausus* to *homines aperti* is relevant to a further issue: the need for a more adequate account of the contacts, exchanges and fusions of different civilisations. The predominance of the closed image of the human being has obviously impeded the recognition and understanding of such processes. William McNeill's critique of Elias (formulated in discussions at the Bielefeld conference) focuses on this problem; as he sees it, neither the original analysis of the civilising process nor later ventures into other areas have done justice to the phenomena of inter-civilizational contacts (including both cultural borrowing and the more autonomous initiatives stimulated by contact). But on closer examination, the two approaches seem easy to reconcile. McNeill stresses the diffusion of inventions that enhance power or wealth and the role of civilizational centres that achieve major breakthroughs in this respect. Both these criteria of progress can be interpreted in terms of Elias's 'basic controls', and there is nothing in the latter model that would preclude a stronger emphasis on the interaction between civilisations. On the other hand, the radical transformations of religious consciousness—or, in more general terms, of world views—that McNeill associates with intensified contacts between cultures and the need to account for a more variegated experience obviously involve what Elias calls 'means of orientation', in contrast to the means of control in the more specific sense. A study of intercultural encounters and their repercussions would thus be particularly useful for the clarification of this distinction.

[29] Elias, *Process of Civilisation*, pp. 262–3.

eighteenth century,[30] Elias's revision of traditional historiography goes even further: he relativizes not only the innovations of the early modern era, but also the two revolutions. I have already referred to his views on the industrial revolution (see above, Section 2); in the case of the democratic revolution, the attempt to minimise discontinuities is even more pronounced. The emergence of a 'democratic regime' is seen as the natural long-term consequence of centralisation and monopolisation; as the apparatuses of power become more complex, they inevitably enter 'the phase in which control over the centralised and monopolised resources tends to pass from the hands of an individual to those of ever greater numbers, and finally to become a function of the interdependent human web as a whole, the phase in which a relatively "private" monopoly becomes a "public" one'.[31]

Of all theories concerned with the rise of the West, Elias's version is probably the most sceptical of the notion of modernity as a new beginning. And there is no denying that his emphasis on continuity throws new light on some central problems. To mention only the most important one, the general minimisation of ruptures entails a corresponding relativisation of contrasts. In particular, Elias stresses complementary rather than antagonistic relationships between the structures and processes of the monetary economy and those of the feudal world. The emergence of a capitalist economy and the emergence of the absolutist state are two closely interrelated and mutually reinforcing sides of the same process. With regard to the latter, the dependence of the two monopolies—the military and the fiscal—on a monetary economy is obvious. With regard to the former, Elias's argument is much less systematic, but it clearly points in the same direction as Braudel's later work: capitalism is not simply a generalised market economy; in a historical perspective, it should rather be seen as the product of structures of domination—including states and state systems—interfering with structures of exchange.[32] This double background remains decisive throughout the capitalist epoch; Marx's theory of primitive

[30] Jacques Le Goff, *Time, Work and Culture in the Middle Ages* (Chicago: University of Chicago Press, 1980).

[31] Elias, *Process of Civilisation*, p. 310.

[32] Fernand Braudel, *Civilisation and Capitalism, 15th–18th Century*, vol. I, *The structures of everyday life: the limits of the Possible* (London: Collins, 1981). A more radical—and much more speculative—variation on the same theme is Immanuel Wallerstein's suggestion that 'the creation of historical capitalism as a social system dramatically reversed a trend that the upper strata (that is, the ruling classes of feudal Europe, J.P.A.) feared, and established in its

accumulation recognised its importance in the formative phase, but this insight was overshadowed by a tendency to equate capitalism with a generalised commodity production that became even stronger in the later Marxist tradition.

In comparison with Marxist and mainstream theories of modernisation, the conception of a parallel and interconnected, but not necessarily coordinated development of capitalist and non-capitalist structures has some obvious merits. Among other things, it suggests a more balanced perspective (not explicitly spelled out by Elias, but quite compatible with his main theses) on the demise of European hegemony in the twentieth century. The catastrophe of 1914 was neither the result of a pure logic of capitalist development nor—as argued in a recent critique of traditional views—brought about by the 'persistence of the old regime';[33] its causes must be sought in the combined dynamics of the capitalist world system and the military-bureaucratic empires that had been drawn into the overall transformation and developed their own lopsided, but by no means inconsequential patterns of modernisation.

However, the discovery of previously overlooked continuities and complementarities does not obliterate the other side of European history; rather, it underscores the need for a more adequate interpretation of what Horkheimer called the 'tradition of upheaval' (*Tradition der Umwälzung*). One of the seminal but underdeveloped themes in Max Weber's comparative analysis of civilisations is the contrast between the Western sequence of revolutions, beginning with the constitution of autonomous urban communities in the Middle Ages, and the absence of anything comparable in the major non-Western traditions. The most emphatic version of this idea in contemporary social theory is Castoriadis's distinction between heteronomous societies and societies with a built-in vision of autonomy. As he sees it, the most profoundly original feature of Europe as a cultural area is the repeated emergence—first in the Greek city-states, and then again (in a significantly different, but not unqualifiedly superior version) in the mediaeval and modern West, from the urban social movements onwards—of a societal project that varies enormously in scope and consistency, but remains nevertheless linked to the perspective of an autonomous society and an autonomous individual, meaning an explicit

place one that served their interests even better' (*Historical Capitalism* (London; Verso, 1983), p. 43).

[33] Arno J. Mayer, *The Persistence of the Old Regime* (London: Croom Helm, 1981).

self-institutionalisation of the former that presupposes the conscious participation of the latter.[34] An important corollary of this thesis is the critique of traditional—primarily Marxist—interpretations that have identified the driving force behind this transformative movement with the dynamics of capital, its protagonists with social classes and its main landmarks with political revolutions. As I have argued, Elias's theory of the civilising process lends some indirect support to his critique; it can be read as an attempt (not the only one, but one of the first and most sustained) to reduce the three above-mentioned factors of discontinuity to surface manifestations of an underlying and unbroken trend.

But the convergence—if we can call it that—is only a partial and negative one. Elias's analysis goes beyond the relativisation of apparent ruptures to cast doubt on the very idea of a 'tradition of upheaval'. For all the emphasis on the social ascent of urban and working classes, social movements—as culturally defined projects of collective action—are conspicuously absent from his developmental model; and the democratic revolution is reduced to an immanent and in the long run inevitable transformation of the power structure.[35] A systematic argument to the contrary would have to take into account a whole complex of economic, political and cultural determinants; here I shall only suggest that the cultural presuppositions of democracy—involving a radically new cultural definition of power, as well as the relativisation and reconstitution of collective identities—are a particularly important aspect. The outlines of this problematic are easily identifiable on the basis of current debates. Some contemporary theorists

[34] See Cornelius Castoriadis, *The Imaginary Institution of Society* (Cambridge: Polity, 1987).

[35] Appendix XVI in Elias, *Process of Civilisation*, pp. 585–6 (originally a long footnote to p. 267) contains more allusions to the changing image of power in an increasingly complex society. Elaborating on his statement that 'the law is a function and symbol of the social structure or—what comes to the same thing—the balance of social power', Elias argues that as the 'direct visibility' of power is replaced by more abstract networks of interdependence, 'A so-called general law—that is, a law applicable and valid equally over the whole area for all the people within it' must develop. However, it is hard to see how the intrinsic dynamics of a more impersonal power structure could account for the whole range of metamorphoses and polarisations characteristic of the modern era. Elias's model is perhaps capable of explaining the secular trend of bureaucratisation as well as the interplay of centralising and diffusive tendencies within this process, but not—to mention only the most obvious problem—the contrast between democratic and totalitarian transformations of the relationship between power and society. Nor is the functional adjustment of law to a more complex 'social balance of power' a sufficient reason for the tension between normative grounding and purposive rationalisation in the development of modern legal structures.

stress the significance of universalistic principles of theoretical and practical rationality, manifesting themselves in a permanent critical distance from pre-given traditions and normative structures (Habermas); of the difference between *société instituante* and *société instituée*, clarified and activated through the questioning of institutional frameworks and the creation of new ones (Castoriadis); or of a new interpretation of power, representing it as a *lieu vide*, essential to the self-constitution of society, but never fully appropriated, completely instrumentalised or adequately incarnated by an actor or a collectivity (Lefort). Such approaches are certainly neither reducible to a common denominator nor reconcilable within a synthesising approach, but their thematic affinities are obvious.[36]

An increased emphasis on culture as a transformative factor also entails a reconsideration of its relationship with capitalism. Here Braudel's work seems to open up an interesting line of research; cryptic and inconclusive as his remarks on culture are, they nevertheless amount to a clear-cut reversal of the traditional approach that has generally been formulated in Weberian terms (although not without some simplification of Weber's original conceptual scheme): it is not so much the emergence of a 'spirit of capitalism' from the internal dialectic of culture that we should try to explain, but rather the ability of capitalism to absorb cultural resources and transform cultural contestation into a more reliable and dynamic cultural defence of the dominant order.[37]

It might be objected that these problems are outside the scope of Elias's work, and that his concrete analysis of the 'sociogenesis of the state' is not inseparably linked to the more tentative overall scheme. Even this qualified defence, however, would come up against some difficulties; to conclude the discussion, I shall briefly indicate some key points in the analysis of state formation where the need for a supplementary 'culturological' account is particularly noticeable.

(a) In Elias's analysis, the struggle between the empire and the papacy in the High Middle Ages does not appear as a significant episode in the 'sociogenesis of the state'; the Church is reduced to one of many actors on the feudal scene, and the imperial project to a premature and overextended

[36] For Claude Lefort's position, see particularly, *Les formes de l'histoire: essais d'anthropologie politique* (Paris: Gallimard, 1979), and *L'invention democratique* (Paris: Fayard, 1981).

[37] Braudel, *Civilisation materielle, economie et capitalisme*, 3 vols (Paris: Armand Colin, 1979), vol. I, pp. 537–48. [English translation, *Civilization and Capitalism*, 3 vols (London: Collins, 1981–84)—eds.].

attempt at territorial unification that later gave way to state formation on a much smaller scale, lagging behind the English and French models (in keeping with this general downgrading, Frederick II is not even mentioned in the genealogy of court society). Yet it would seem that the specific impact of this conflict on European history was in large measure due to the clash and mutual undermining of two different cultural interpretations of power. The failure of both the imperial and the papal bid for hegemony ruled out some conceivable alternatives that are best described by analogy with other civilisations (at one extreme, we can think of Byzantine caesaropapism, at the other of a unified Christendom where the position of the clergy might have been comparable to that of the Confucian literati in China; the first perspective is a commonplace in modern historiography, whereas the second was, if I am not mistaken, first suggested by William McNeill); it opened the way for the more successful consolidation of proto-national monarchies, and at the same time it influenced the context in which this development took place; finally, the struggle was accompanied by a mobilisation of intellectual resources that had profound and long-lasting consequences.

(b) Although Elias's analysis of the absolutist state is perhaps the most incisive and solidly documented part of his theory, it leaves open some questions concerning the cultural model of absolutism. On the ideological as well as on the institutional level, the absolutist paradigm fuses traditional and modern elements in a synthesis that transforms both components and imparts its distinctive stamp to a whole historical epoch. Differences in the details and proportions of this combination would be one of the main themes of a comparative study of absolutism. An especially interesting case in point would seem to be the contrast between the French and the Spanish model; the significance of the latter is enhanced by its rôle as the cultural basis of the Catholic and absolutist counter-offensive in Central Europe in the seventeenth century. The Spanish line of development, however, is left out of account in Elias's model; Spain is not even mentioned in the 'survey of courtly society' at the beginning of the second volume of *The Process of Civilisation*. If it is—up to a point—legitimate to speak about an 'alliance of the courts with the humanist culture and with the scientific movement',[38] a comparison of the French case with others would help to specify the preconditions for this alliance to become relatively stable and effective.

[38] Christopher Dawson, *Understanding Europe* (London: Sheed & Ward, 1953), p. 43.

The antecedents of the absolutist synthesis can be traced back to the Middle Ages. As Georges Duby has shown,[39] two different images of power begin to take shape in fourteenth century Europe: on the one hand, a notion of kingship as the incarnation of sovereignty, on the other a more 'civic' conception, closely linked to the revival of Roman law and tending towards a more abstract notion of the political sphere. The first points towards the doctrine of the 'divine right of kings'; it presupposes a relative secularisation of Christian values and a demarcation from the imperial visions of the Middle Ages. The second is the first step towards the understanding of the state as 'an omnipotent yet impersonal power'[40] that stands at the end of the formative phase of modern political thought. The absolutist state rests on this double foundation.

(c) Finally, a few words should be said about the question of nationalism and its role in the process of state formation. For Elias, this is obviously a very marginal factor.[41] When he tries to subsume the conflicts between modern states and those between the much more fragmented power units of the early feudal world under a general model of competition and monopolisation, he implicitly minimises the specific characteristics of national states and their interrelations.[42] His theory of the civilising process contains no distinctive interpretation of the nation as a pattern of collective identity or of nationalism as a socio-cultural current. In this respect, his approach is a particularly clear-cut version of the generally reductionist stance that still prevails in sociological theory: nationalism is mostly treated as a form of social integration and communication that corresponds to the needs of a modern—that, highly differentiated, mobilised, and secularised—society.[43] The concept of 'nation-building' epitomises this functionalist view. On the other hand, critical reactions against it often shift the problem to the psychological level. Anthony Giddens, for

[39] Georges Duby, *Le Temps des cathédrales: l'art et la société 980–1420* (Paris: Gallimard, 1976), pp. 296–327. [English translation, *The Age of Cathedrals* (London: Croom Helm, 1981)—eds.].

[40] Quentin Skinner, *The Foundations of Modern Political Thought*, vol. 2 (Cambridge: Cambridge University Press, 1978), p. 358.

[41] Elias's later book on German nationalism and the German national habitus was first published in the same year as Arnason's paper (1989). See Norbert Elias, *Studies on the Germans: Power Struggles and the Development of Habitus in the Nineteenth and Twentieth Centuries* (Dublin: UCD Press, 2013 [Collected Works, vol. 11]).—eds.

[42] See particularly Elias, *Process of Civilisation*, p. 267.

[43] See Ernest Gellner, 'Nationalism', *Theory and Society*, 10: 6 (1981), pp. 753–76, for a succinct and up-to-date analysis of nationalism from this point. of view.

example, insists on the distinction between the nation-state as an institutional phenomenon, based on the unification of an administrative apparatus, and nationalism as a psychological phenomenon, 'involving felt needs and dispositions' and responding to transformations of everyday life that undermine the 'ontological security of the individual'.[44] Both the functionalist and the psychological approach neglect the 'culturological' problematic of nationalism as a new type of collective identity and its implications for the political constitution of society; so far, this issue seems to be receiving more attention from historians than from sociologists.[45]

These criticisms should not obscure the merits of Elias's work on state formation. If the emergence of the absolutist state and its transformation into the nation-state is now increasingly seen as not a background, but also a co-determinant of these aspects of modernity that social theorists have mostly taken more interest in (capitalism, industrialisation and democracy, to mention only the dominant themes), *The Civilizing Process* should also be recognised as the single most important preparatory step in this direction.[46] But as I have tried to show, it can also serve to illustrate the shortcomings of a more general perspective shared by many otherwise different theorists: the disproportionate emphasis on power that leads to a neglect of its cultural contexts.

[44] Anthony Giddens, *A Contemporary Critique of Historical Materialism* (London: Macmillan, 1981).

[45] See the recent and important work by John A. Armstrong, *Nations Before Nationalism* (Chapel Hill, NC: University of North Carolina Press, 1982).

[46] The most representative recent work on these problems is Anthony Giddens, *The Nation-State and Violence* (Cambridge: Polity, 1985).

Discussion of Arnason's Paper

Transcript

Dunning: Can I ask a question? I wonder if you would elaborate because I can't see it at all. I mean this idea of Elias's view of development of the science as teleological. I mean, it's to be looked beyond that, and see no element of teleology in it whatsoever.

Arnason: Teleological in the sense that it's directional. And let me make it clear, I'm referring most specifically to his programmatic essay on the sociology of knowledge, the one that was published in *Sociology* in two instalments, 'Sociology of Knowledge: New Perspectives'. Teleological in the perhaps loose sense that, in contrast to his view of social process, as far as I can see here, there is an overall trend towards a final state. The logical cognitive progress is the gradual approximation to objective reality, is the increasing *Realitätsnähe* and *Objektadäquanz* of objective knowledge. There is nothing like that in Elias's view of social process. He is at pains and he has some very strong arguments in that (concept) to eliminate all connotations of an overall evolutionary direction of social process. This is what I meant.

Transcript (✉)
Berlin, Germany
e-mail: Artur.Bogner@web.de

A. Bogner, S. Mennell (eds.), *Civilisations, Civilising Processes and Modernity – A Debate*, Palgrave Studies on Norbert Elias,
https://doi.org/10.1007/978-3-030-80379-7_13

Bogner: But this is to say that the idea of scientific progress in itself is teleological. You equate directional, the directionalism, the directional character of the idea of scientific progress itself with teleology.

Arnason: No, not the directional character itself, but the definition of a direction in terms of an approximation to a final state—in terms of the approximation to an ideal definable beforehand.

Wallerstein: In which case, however, the alternative is, is it not, is that you can't use the word 'progress' about it? I mean, you may be right about Elias, but I think, maybe Bogner is also right. I mean if we're not directional, then, and we're not approximating a final state, then in what sense is the new scientific knowledge progress over the previous scientific knowledge? Other than approaching to a steady state of truth.

Mennell: Well, I was going to make the same point that when Dr Arnason first replied he appeared to be equating directionality with teleology. And, of course, (another) point of Professor Elias's theorising about social process is to show how directionality can be other than teleological. But I think that this point about Elias's somewhat provocative use of the term 'object adequacy' is the thing that leads to the what I think is a misunderstanding, that I think it might be more appropriate to let Norbert Elias himself explain what he means by 'object adequacy' than for us to pick away at Dr Arnason on the subject.

Elwert: I admire the way you put order into a sociological tradition. It's not only wonderful. It's also convincing. But I had a problem understanding one of the arguments because it was presented not only in the summary, but also in the paper in a very short version. That is the argument on world view, cultural models and the images and phantasms of power. I think if empirically there's a demonstrating, a demonstration of long-term processes (which) transcend changes in world view it is not easy to argue for a re-vitalisation of the old world view sociology. Of course, one might argue that any system of power and any system of critique of power is a cultural model. So what? This would be tautological. This is nothing surprising. So, certainly you mean something else, insisting on the need to bring world view analysis together with this type of sociology of power. Could you explain or give to a simple-minded person like me some examples which could illustrate the argument?

Arnason: That's a different point already of course. But I'll do the first one firstly. The question concerns … well, there are two different questions. I think. The first concerned the legitimacy of the expression 'teleological' and then was Professor Wallerstein's question, granted that it is teleological, but then is it, in any way, possible to defend the notion of scientific progress once you dispense with the idea of approximation to reality. Well, let me start with the observation. I don't think there is a good theory of scientific progress or, more generally, of cognitive progress. I suppose all (here) would agree that the orthodox Marxist or rather the Leninist theory of progress in terms of the infinite approximation to absolute truth is hopeless. Then there is the Popperian one of course which has come under heavy fire. And the criticism that has been levelled against Popper, as far as I can see, is mainly of two kinds. Firstly, that he disregards the role and the discontinuity of interpretative frameworks in which empirical knowledge is embedded. Secondly that in order to save his model he has to reintroduce, through the back door as it were, the idea, the correspondence theory of truth in, again in the form of the idea of gradual approximation. The only, at present the only alternative theory of cognitive progress, the only one that tries to do without the idea of gradual approximation to reality and its foundation in the correspondence ..., the only theory that tries to do without any version of the correspondence theory of truth and the related idea of gradual approximation to reality is that of Habermas, I think, the theory of cognitive progress as the rationalisation and institutionalisation of discourse, cognitive progress as progress towards unconstrained argumentation ...

Wallerstein: Civility.

Arnason: Yes. I think the weaknesses of that theory have been pointed out as well. In the recent reply to his critics Habermas has admitted as much. In other words, I don't think there is a theory that has successfully integrated the three aspects that are at stake here. There is firstly, of course, the referential dimension of knowledge, *Objektbezug* rather than *Objektadäquanz* to use a more neutral expression. There is secondly the embeddedness of this in interpretive frameworks that are not reducible to a continuous line of development. And there is thirdly the dimension of discourse, the embeddedness in, as Habermas would call it, communicative action. To come back to your question: can you, given these premises, talk about scientific progress at all? Not in the sense, I think, of an overall trend towards an ideal state. I think you can only talk about localised (assured), episodic, ultimately fragmentary progresses, rather than progress with a capital 'P'. At least that is the position I would be tempted to retreat to. That is not, to say on that (...) to come back to a question raised by Georg Elwert which has some connection with this one. His argument, if I understand it correctly, is that, assuming that it has been demonstrated, there are such things as long-term directional, but unplanned social processes that undercut the level of conscious interpretations, cultural models, what is the justification for reintroducing, as you call it, the old sociology of world views? But as you could add an implication of a question that you did not spell out, I think, is there any other way of doing it, then? If you want to reintroduce a problematic of world views, is there any other way of doing it than, what Habermas has been doing, to postulate a universal logic of development? World views ultimately rooted in universals of human language. Well, I think there is and I think that is not a necessary implication. The question concerns precisely the status of the long-term directional unplanned, unconscious processes. The question concerns their determinacy or their ambivalence. If you can show that they are ambivalent, if you can say that they are pluriperspectivistic—to use a term that Elias himself has used in describing the structure of social reality. If you can show, to take a concrete example, when the democratic revolution in the broad and long-term sense that is a specific characteristic of the West, that it is not an automatic outcome of shifting balances of power and more complex interdependences, but that it represents the actualisation, partial, interrupted, incomplete, as we will all agree, but still a certain realisation of one possibility, one perspective inherent in the long-term process. Then you are back to asking, why this choice, why this alternative among others and in particular what are, in our present context, what are the cultural presuppositions of this? What are the cultural definitions and images of society that prompted the major rupture of traditional societal patterns that occurred in the West?

Bogner: Perhaps I myself have a tendency in the concern with the theory of knowledge to argue for a retreat on the notion of control in the technological sense. That means you can have a diagnosis of long-term progress in natural sciences in the sense that this process leads, continuously, more or less continuously, to a growing ability of men to control natural processes. And I think the problems you have with this conceptualisation of scientific progress are the problems associated with the philosophical problem of what truth really is. But I think this is irrelevant to a sociological theory of societal development and of the social development of knowledge. It is irrelevant what kind of truth this is. Is it a true truth or only the truth you need to handle, to operate, to manipulate natural processes? For the idea of technological progress, of scientific progress in the sense which is relevant for Elias's theory it is enough if you can diagnose, if you can see and argue for a progress in terms of controllability of nature.

Dunning: Presumably, the end state towards which that is tending is man's ability to create his own universe.

Bogner: Yes, something like this.

Arnason: This could mean two things. Either you are arguing (committed) progress means simply that we can increase our control of nature. That's all there is to it and nothing more. Now if you mean that, and that's an even more radical reduction of knowledge to power than what I was attributing to Elias because then ...

Bogner: No, I only mean this is the dimension in which you can correctly argue for a directional character of the process of growth of knowledge. This is the aspect, there are other aspects too, but I think for a diagnosis of the directional character of this process it is enough and, perhaps, it is the only aspect to which we can ascribe a directional development, a character in that sense.

Arnason: OK, but then two points. I mean, firstly, if you define it like that, if you argue the directional process here is simply the increased control over nature, then firstly, the idea of the end state which you formulated, the idea of an end state that man [*sic*] can create his own universe or, perhaps somewhat less provocatively formulated, the idea of an unlimited mastery of man over nature. And the idea has been conceived and has been defended. It has among other things played quite a role in the Marxist tradition. You find it in exactly those words 'unlimited mastery of man over nature', you find it in Rosa Luxemburg who otherwise was not the most pragmatical Marxist. I think we would all agree ...

Bogner: You find it in Horkheimer and Adorno although ...

Arnason: Yes, but with other connotations.

Bogner: ... although with other connotations.

Arnason: But I think we would all agree in light of later experiences that the very idea is incoherent and self-destructive—the very idea of unlimited mastery of man over nature. So the ...

Bogner: But self-destructive not in a cognitive sense when it is a wrong conception.

Arnason: Your point was that there is no cognitive sense. There is only the sense of mastery, control. There is no such thing as an independent cognitive sense because you are not renewing the Engelsian argument. The Engelsian argument is: the proof of the pudding is in the eating. The fact that we can control nature more and more effectively proves that our knowledge of it reflects the objective state of things more and more adequately. You are not renewing that.

Bogner: No, I am not taking this stance.

Arnason: But then the cognitive dimension disappears and what remains is simply and solely the power dimension.

Elias: It is a bit difficult for me to answer it and I've listened with great attention. I have made ample notes and I have learned from what you have been saying, it will give me material for further thought. But it may sound slightly ludicrous if I say, I am still in the middle of my work. And there is a good deal to come. Among others, some progress with regard to the very question which you just have raised. I can only say that this I can do only by having a clearer idea as to what knowledge is. At present, theories of knowledge have no clear statements about what, in human terms, knowledge is or how it functions.

Arnason: I wholly agree.

Elias: And unless that has been done, the question which you raised will remain unanswerable. I will do that, and I have done it in terms of a symbol theory of knowledge. But that is yet to come and I feel, at the moment, I cannot say very much more than to say I think it becomes quite clear, the relationship between an unplanned process, which the knowledge process definitely is—it's an unplanned process—and what you call teleological. I personally think what I have done is as little teleological as all my other works, although it contains the concept of an advance, an advance in a certain direction, but with no final aim or end. Because, if I may put it in a slightly ludicrous manner, the world is infinite and I do not think that we shall ever—as all symbols, all knowledge is interrelated—come to the end of it. I do not see a final state. And that is also the difference. I mean, and that you will I think understand, it's simply the old vocabulary, 'truth', which assumes a final state, does not take account of the interrelatedness of the whole universe. In some respects we're approaching nearer, but there is no final state in which, if I may say so, our symbols are, as it were, adequate symbols of the universe. So, let me attack the problem from another side. I would like to give you a problem which was very much in my mind when I listened to you. I can perhaps best express it by means of a simile. I am not sure that I find it particularly fruitful only to analyse concepts without ever making a reference to empirical work for which they can be used. All the concepts and all the theoretical work I am doing are really forging instruments or making, giving them a stronger edge so that we can investigate more fully the real societies which exist. So a mere brooding or analysing or clarifying of concepts is a little bit, so it seems to me, as if you wanted, to use a very well-known simile, if you want to teach a person swimming, you cannot, on the land, show him all the movements. In the end, unless you throw him or her into the water, he cannot swim. And the same is true with the concepts which you have very neatly and very effectively analysed. What do they mean if we want to swim? So I would, for instance, say I found, simply, in your analysis an over-accentuation of certain concepts and an underdevelopment of certain other concepts. For instance, you have not said very much on what distinguishes my whole approach from others except Marx is that it is very concentrated on processes. That is, unless one learns to think in terms of processes and stages of development, unless one can use a dynamic approach in practice, all the concepts like 'system' and 'action' remain really useless. It is only in the context of a process context that all the concepts which you have used come really to life. And there I do not really, I cannot … I mean you have made very many points. I have made a good deal of notes for them. And you'll forgive me if I do not take up each of them. But I want to say one general thing. And that is I think that the social development altogether is, broadly speaking, an unplanned development. We have made plans. They are short-term plans, but the intertwining of plans, as you probably know, creates again and again an unplanned development. Now, I do not make the mistake which many sociologists have made in the past and are still making today. I do not assume that the blind social processes go necessarily in the direction of my wishes or my ideals. I do not think that they go necessarily in the opposite direction. But I do see that there are many possible directions into which they can go. Neither of them necessarily coincides with my wishes. But I do think, and now of course, if you want, comes my bias. I do think that a better knowledge of the functioning of the blind processes may eventually enable human beings to steer their fate, their social fate, better than they can do it today. At the moment, if I look at politicians, who direct our fate, my God, there's so much dilettantism, so little knowledge of how to handle social processes. But I think the sociologists and the historians have a very good task in simply trying to give more object-adequate, more reality-congruent models of the reality. Not that they should become politicians. You may not have yet read one of my latest things about power and knowledge in which I have introduced a little utopia showing exactly the limits which I set for sociologists. They are not politicians; they have to do their spadework in investigating, as best they can, how things are. And the politicians may be using it, but they must not themselves become politicians. I do not think that one can deny progress in the natural sciences. If one has a concept of development, one has to work with breakthroughs to a new level constantly. You have to see it in the biological evolution where you come to a different stage where suddenly there is a breakthrough, (to) a new level where living organisms can do things which no living organism could do before. So one has to work with a different theoretical apparatus in discussing what we call 'developments'. And what is most lacking I think in sociology is a really … is a concept of development with which one can work, with which one can, for instance, say what breakthrough, how the breakthrough to what we call the natural sciences was possible. That has not yet been established. Or, how the breakthrough to living in towns was possible. Really, it seems to me that historians do not sufficiently clearly define what distinguishes their periods from each other. In sociological terms, it is not enough to say 'Renaissance' but there has to be a very clear statement of what the breakthrough was which has been represented, shall we say, by classical antiquity, by, what was the change which occurred from … why do we talk of the 'Middle Ages'? And there I myself have, am going to suggest a new period arrangement. I would think one should speak of the 'earlier antiquity', which was clearly, where knowledge was clearly dominated (by) and the possession of priests, one can distinguish that very clearly from a later antiquity in which knowledge became secularised. And one can distinguish that, (in) part Keith Hopkins has shown it today, from the Middle Ages which surely begins when Constantine makes Christianity a state religion. When knowledge again becomes dominated by priests, but *not* on the old level. It is quite clear that what happens now, when the Church becomes state church, is something which is not going back to the old level of priest-dominated knowledge, but there is something new. So, our concept of development is far too simple by having a clear-cut progression. You have, for instance, the absorption of the secular knowledge of antiquity into the knowledge of the Middle Ages. The theology, it is not, as it is usually said, that philosophy is a daughter of theology. It is also that theology is a daughter of philosophy. Without the antecedent philosophy, what we call theology would not have been possible. Certain concepts, there is a direct continuity of knowledge from the old philosophy to the new theology. For instance, concepts like 'cause' are not lost, (they) are taken over into the new theology, and one has now to use them in terms of, shall we say, *causa causans* where God becomes the *causa causans*. But my problem is that without the taking over of the secular knowledge of the later antiquity into the early Middle Ages, probably the new secularisation in the Renaissance would not have been possible. I am quite convinced that without the transmission of the Latin language and the Latin knowledge, the continuity of knowledge in, through the Church, that would not have been possible. So let me just say one word more. I think we have then, what I would call the 'Later Middle Ages', what we call 'modern times', which in my view are the later Middle Ages, when a new wave of secularisation comes and when certain things, for instance like wars, are still as unmanageable as they were before. So I think we should see that what is most lacking in our—one of the things which is most lacking in our present knowledge is a very clear concept of development. And it is here and, if I may say so to Professor McNeill, I think that there is one of the main differences between a sociological approach and a historical approach. ... And yet, it is, for instance if one looks at human knowledge, not to be denied that the fund of our knowledge has grown, that it has developed, that it has a directional character. But it is a blind character. And I do not think at all that one can give it a final aim or that there is a final goal. We have to get used to the fact that there are observable directional developments going through a series of clearly articulated stages, always capable of being reversed, as they were when the West Roman Empire broke down. So our whole civilisation, our whole societies can, as it were, break down and the whole processes can be reversed. But I do think that the main differences between what—I mean I have taken a great deal of inspiration from the work of Professor McNeill. But I do think that this concept of development which I tried, to some extent, to illustrate in my talk which I gave before, is lacking. And perhaps, I do not know that what I have said is not at all, does no justice to what you have said. I have not entered all the points which you made. But perhaps what I have said goes some way towards answering some of your problems.

Arnason: When Professor Elias says there is no theory available or in sight that could give us a satisfactory account to what knowledge is, what it means in human terms, I completely agree. In fact, I think that converges with the rest of my argument. But perhaps about all we can say on the basis of contemporary theories of knowledge is that we can identify the task of such a theory of knowledge in terms perhaps of three points. What it would have to do is to bring together the referential, the interpretive and the discursive dimension of knowledge. By referential aspect I mean the openness to experience, by interpretive aspect I mean the embeddedness in interpretive frameworks that are neither reflections of reality nor stages on our way to an ... and by discursive aspect I mean the embeddedness in what Habermas and Apel call *Argumentationsgemeinschaften*. But then, if that is so, if there is this huge agenda in front of the theory of knowledge, that only confirms my point I think, you cannot reduce the problem to control of nature. Very briefly, concerning the other points, I thought I had taken into account and explicitly recognised that I am dealing with a work in progress and I am even more conscious of that after I listened to Professor Elias's contribution yesterday which did add some significant new elements. As for the underdevelopment of the historical part versus the conceptual, I confess to a certain privilege of conceptual analysis but in the last part of the paper, I did make an attempt to shift the argument to a more concrete historical ground. I mean, I do take up some concrete problems of (West) European history, the democratic revolution, the absolutist state, the role of nationalism, such concrete cases as the Copernican revolution. I think perhaps (I should) explain a bit more why I discussed the case of the Copernican revolution at such length. The point is that not only is there no theory of knowledge and no satisfactory interpretation of cognitive progress. But there is also, there is no paradigm, there is no example of an unequivocal breakthrough to objective knowledge. The Copernican revolution is about the most likely candidate and my point was that it's a very complex and ambivalent phenomenon. And that's about all.

The Domestication of Fire as a Civilising Process

Johan Goudsblom

Part One: The First Stage—Origins and Conditions

Introduction

The ability to handle fire is, along with language and the use of tools, one of the basic acquisitions of all human societies from prehistoric times. Moreover, to an even greater degree than either language or the use of tools the control of fire is not only a *general* but also an *exclusive* human acquisition. Rudimentary forms of language and tools can also be found among primates and other animals; 'man',[1] however, is the only species which has learned to handle fire.

[1] The concept of 'man' is used here to designate the species known as *Homo* in biology and palaeoanthropology including *Homo habilis*, *Homo erectus*, and *Homo sapiens* (see David Pilbeam, 'The descent of hominoids and hominids', *Scientific American*, March 1984, pp. 60–9). When referring to human social life in a sociological sense I prefer the plural form

J. Goudsblom (✉)
Amsterdam, Netherlands

A. Bogner, S. Mennell (eds.), *Civilisations, Civilising Processes and Modernity – A Debate*, Palgrave Studies on Norbert Elias,
https://doi.org/10.1007/978-3-030-80379-7_14

It is customary to distinguish two big transformations in the history of human civilisation: the 'agrarian revolution', marked by the domestication of plants and animals, and the 'industrial revolution', marked by the intensive utilisation of inanimate energy. This view somehow implies that 'civilisation' began only after some societies had entered the first stages of agriculture.[2] However, the agrarian and industrial transformations had been preceded, many hundreds of thousands of years earlier, by the transition to the active use of fire—a transformation that was of at least the same magnitude and without which neither agrarianisation nor industrialisation could have taken place.

In an analogy to the domestication of plants and animals, we can also speak of the far more ancient domestication of fire. In the course of this long-term process, what was initially a wild, incalculable and dangerous natural force has been tamed to a certain degree, and incorporated into human society. We are dealing here with a development extending over many hundreds of millennia, from long before the emergence of *Homo sapiens*, which is closely linked with the process of humanisation or the formation of humanity in its present shape. It has had profound effects both upon the relationships of humans with all other animals who have not learnt the art of mastering fire, and upon the mutual relationships between human beings. There is every reason to assume that these changes have not stopped at the level of the individual personality.

Stories relating how 'man' has come to obtain fire have a prominent place in the mythologies of many peoples. These stories show how long people all over the world have continued to consider fire as something very special to which a divine origin is ascribed. Later on, fire has also played an important part in philosophical cosmologies and in early natural science. Anthropologists from Edward Tylor to Ralph Linton have written at length about the control of fire and its great importance for the

of 'humans' or 'people'. See Johan Goudsblom, *Sociology in the Balance* (Oxford: Basil Blackwell, 1977), pp. 126–31.

[2] For some representatives of this view, see the articles on the 'Urban revolution' in the *International Encyclopedia of the Social Sciences* (1968), vol. 16, pp. 201–21.

development of humanity.[3] In contemporary social science, however, the subject has increasingly disappeared from sight.[4]

The following essay is part of an attempt to trace the main lines in the domestication of fire. In order to assess the implication of the mastery of fire—both its precondition and its effects—I shall begin by considering the first stage, when humans were not yet capable of handling fire. By going back to this very first stage I hope to be able to indicate certain basic relationships of dependence and power which continue to enter into the control of fire to the present day but which are often so much taken for granted that they are hardly noticed any more.

In subsequent work[5] I plan to examine how the further development of fire control is related to the next great transformations of human existence ('mutations' as Linton once called them[6]), the domestication of plants and animals, and the domestication of mechanical energy, more commonly known as the agricultural and the industrial revolutions. There too I shall be primarily concerned with the question of what were the conditions for new spurts in the development of fire control and what consequences these developments have had for human society.

The scope of the entire project is very large. The action (to speak in terms of the rules of classical drama) concerns the human use of fire, the place where the action takes place is the earth, and the time extends to at least half a million years. In order to keep this vast subject within focus I

[3] For myths about fire see James George Frazer, *Myths of the Origin of Fire* (London: Macmillan, 1930); for cosmology and natural science, Gaston Bachelard, *La Psychanalyse du feu* (Paris: Gallimard, 1937); for anthropology, Edward Burnett Tylor, *Researches into the Early History of Mankind and the Development of Civilization* (London: John Murray, 1878) and *Anthropology: An Introduction to the Study of Man and Civilization* (London: Macmillan, 1881), and Ralph Linton, *The Tree of Culture* (New York: Knopf, 1955).

[4] A telling indication of the diminishing interest taken by the modern social sciences in the subject of fire and fire control is the fact that the *International Encyclopedia of the Social Sciences* of 1968 does not contain the word 'fire', neither as an entry nor even in the general index. It is also noteworthy that Charles Lumsden and Edward Wilson in their book with the promising title *Promethean Fire: Reflections on the Origin of Mind* (Cambridge, MA: Harvard University Press, 1983) nowhere refer to actual fire or the use of fire.

[5] This conference paper was presented at a fairly early stage in Goudsblom's research; a slightly revised version was published in *Theory, Culture and Society* 4: 2–3 (1987), pp. 457–76, but for his final views on the subject readers should refer to his book *Fire and Civilization* (London: Allen Lane, 1992).—eds.

[6] Linton, Ralph, 'Present world conditions in contemporary perspective', in Ralph Linton (ed.), *The Science of Man in the World Crisis* (New York: Columbia University Press, 1945), pp. 201–21, at p. 213.

have tried to approach it from a sociological point of view, and to consider the control of fire as a function of social interdependence and social–cultural development. The fact that human beings have managed at all to learn to handle fire I regard as an outcome of their living together in societies. The use of fire and the care for it require everywhere certain social arrangements and individual adjustments. Fire exerts inexorable constraints to which people respond by developing commands and prohibitions and forms of self-restraint. The control of fire is always part of a system of social control and self-control; in this respect, its development may be regarded as a 'civilising process' in Norbert Elias's sense of that concept.[7]

The physical effects of fire have not changed during human history. Its direct contact with the skin causes burning wounds almost immediately, at all times. What has changed, however, are the kinds of fire with which people are confronted and the ways in which they react to it. As the human capacity to control fire has increased, people have generally come to depend more and more upon social arrangements regarding fire, such as the supply of fuel and safeguards against fire-risks.

In order to acquaint myself with the many problems related to the domestication of fire I have read into the literature of various fields such as archaeology, palaeoanthropology, ethology and ecology, in addition to social sciences in a stricter sense of the term such as anthropology, psychology and (my own field) sociology. Reading in the professional literature which I have consulted—tops of gigantic icebergs—I have become painfully aware of my limitations as a non-specialist. At the same time, I have noticed how much room there is for connections between the diverse specialisms and how rewarding the challenge is to look for such connections.[8]

[7] See Norbert Elias, *On the Process of Civilisation: Sociogenetic and Psychogenetic Investigations* (Dublin: University College Dublin Press, 2012 [Collected Works, vol. 3]); original publication, *Über den Prozess der Zivilisation*, 1939; earlier English editions published under the title *The Civilizing Process*. As early as 1916 Walter Hough remarked about the use of fire: 'Clearly no civilising influence is greater and much of man's physical modification is doubtless due to the cultural structures growing out of the utilisation of fire and the reciprocal and cumulative effect on habits of life moulded by relationship to fire' (Walter Hough, 'The distribution of Man in relation to the invention of fire-making methods', *American Anthropologist*, n.s. 18 (1916), pp. 257–63, at p. 258).

[8] How tenuous the connections between disciplines often are is evinced by the various datings of the first use of fire. While according to archaeologists and palaeoanthropologists this first use dates back to at least 500,000 years ago, in the ecological literature the range is

Before Domestication

The question to what they owe the control of fire has occupied people of old. In many cultures there are myths relating to how certain people have managed to obtain fire, often by means of a trick, from gods, or from animals who apparently possessed control over it before but had to give up this precious possession to humans. As is characteristic of myths they represent the incorporation of fire into human society as one single event, or as a climax of one coherent set of events.[9]

Actually, as we can surmise, the conquest of fire must have been a long process extending over hundreds of thousands of years, with many small steps forward often followed again by regressions. Ecological research shows that fires originated by lightning, volcanoes, or other natural causes occur with more or less regular intervals in most parts of the land surface of the earth—with the exception of those areas such as polar regions or tropical rain forests which contain little quickly inflammable matter. The animals in a region hit by fire experience it in the way that they undergo rain and snow or heat and cold—as events which come about, which they cannot control in any fashion, and to which they try to adjust for better or for worse. Virtually the only possible adjustment to a blazing bush fire is flight; whoever does not find safety in time, perishes by suffocation, by desiccation, or by burning.[10]

The immediate consequences of a fire seem disastrous to vegetation and wild life. Everything that has been touched by the fire is charred and reduced to ashes. All that remains is a bare and scorched landscape. In the course of time, however, the traces of the fire disappear almost completely. Nature recovers: out of old roots and newly wind-blown seeds fresh plants

sometimes limited to 50,000 or 55,000 years (see John Phillips, 'Effects of fire in forest and savanna ecosystems of sub-Saharan Africa', in T. T. Kozlowski and C. E. Ahlgren (eds), *Fire and Ecosystems* (New York: Academic Press, 1974), pp. 435–82, at p. 440). An ecologist from Israel even states in full earnestness that 'probably the first man-made fire was set by Samson when he used firebrands between the tails of 300 foxes to burn the standing corn, the stooks of corn, the vineyards, and olives of the Philistines' (Z. Naveh, 'Effects of fire in the Mediterranean region', in Kozlowski and Ahlgren (eds), *Fire and Ecosystems*, pp. 401–34, at p. 408, with reference to Judges 15: 5).

[9] Frazer, *Myths of the Origin of Fire*; Claude Lévi-Strauss, *The Raw and the Cooked* (New York: Harper and Row, 1969) and *From Honey to Ashes* (New York: Harper and Row, 1972).

[10] As a way into the sociological literature, I have used T. T. Kozlowski, and C. E. Ahlgren (eds), *Fire and Ecosystems*. For the reactions of animals to wildfire see especially J. F. Bendell, 'Effects of fire on birds and mammals', pp. 73–138.

and bushes shoot up and, as ecologists have found, in the longer run the effects of fire are beneficial, and even essential to many plants. It checks diseases by killing fungi and insects. Many forests and savannahs consist of 'fire climax' vegetation in which the balance between various species is partly maintained by periodic burns.

Especially in forests, these fires are usually of a restricted size. One of the consequences of the more or less regularly occurring fires is that they help to prevent larger fires. If, in a forest, lightning causes a conflagration, this will destroy mostly all the dry wood and accumulations of barren leaves and needles. Subsequently, on the spot, young vegetation will arise, with a great resistance against fire, making a recurrence of fire during the next few years unlikely. Thus many forests in prehistoric times will have known a certain 'fire-cycle', a 'natural burning pattern' which has been disturbed practically everywhere when humans later came to dispose of fire.[11]

Assuming that humans, at the beginning of the Stone Age, did not live in dense rain forests but in more open savannah-like areas, it seems reasonable to posit that fire of old has formed part of their natural environment. In contrast to other animals, human beings have not restricted themselves to a more or less passive adjustment to this natural phenomenon but they have gradually learned to incorporate it in a more active manner into their own societies. In the course of this development, as Frazer and many others have noted, three stages can be roughly distinguished.[12]

1. *No, or passive use of fire.* During this stage, people, depending on where they lived, will more or less regularly have come across naturally caused fires. It is slightly misleading to speak of 'the discovery of fire', for people will always have had encounters with it. Although they may gradually have learned to appreciate that fire sites could be very useful as a finding-spot for food and as a source of heat and light, throughout this stage, fire remained an essentially wild untamed force upon which they had hardly any hold at all.

2. *Collecting and preserving fire.* The actual domestication of fire did not begin until the second stage, during which people learned, no doubt

[11] See Kozlowski and Ahlgren (eds), *Fire and Ecosystems.* The impact of human interference with fire is emphasised in the contributions by Carl O. Sauer, 'The agency of man on Earth', pp. 49–69, and Omer C. Stewart, 'Fire as the first great force employed by Man', pp. 115–33, in William L. Thomas, Jr. (ed.), *Man's Role in Changing the Face of the Earth* (Chicago: Wenner-Gren Foundation, 1956).

[12] Frazer, *Myths of the Origin of Fire*, pp. 210–26,

with many falterings, to use and control fire actively. They managed to pick it up and transport it and to keep it burning for longer periods at spots of their own choosing.

3. *Making fire*. Much later again people will also have learned to make fire themselves so that they would not have to rely exclusively any longer upon already existing fire brands. For a long time either friction, produced by rubbing or rotating wood, or percussion, produced by striking metals and flints, were the only methods available to ignite a fire. Although these methods demanded great skill, they were mastered by at least some members of every known society in recorded history. Only with the advance of ready-made instruments such as matches and lighters, have these skills become obsolete in many places.

In social and cultural development, the characteristics of earlier stages rarely disappear altogether during the later stages, but they do undergo drastic changes in form and function. Thus, fire does not cease to be a wild untamed force during the third stage; in the contemporary world, it can even be far more dangerous precisely because of conditions brought about by people in modern society. In the same way, it remains possible of course for people to 'borrow' fire from existing hearths, even though matches and lighters have made the art of kindling fire exceedingly simple.

The first stage during which there was not yet any active control of fire we can only visualise in a somewhat speculative fashion. There is no ethnological description of any people who have lived without the active use of fire. One of the most suggestive evocations of the stage of passive use of fire is still to be found in the report by the German explorer and anthropologist Karl von den Steinen of his expedition along the Xingu river in the interior of Brazil in 1887–1888. He begins with a lively description of the consequences caused by the camp-fires of his own expedition:

> The fires which we lit during our journey often burned for days and spread spontaneously over large distances. Curious and striking was their influence upon the animal world. All sorts of predators took well considered advantage of the event; they sought and found their victims, not so much by the bright flames, but rather amidst the smouldering ashes in which many a rodent would lie charring. Numerous falcons were hovering over the dark clouds of the *queimada*, game was running to it from afar to lick at the

> salted ashes, preferably at night because they could not hide in the barren plain. The ground radiated a comfortable warmth.[13]

On the basis of these observations, Von den Steinen continues with a brief digression on the lessons which human beings already at a very early stage could have drawn from natural bush fires. At the outbreak of the fire, they would have seen game fleeing. Later they would have basked in the glow of the extinguishing fire, and they would have picked up partly charred animals and fruits from the ashes and savoured them. In this manner they would have learned to appreciate the advantages of broiling and roasting which not only added to the taste of meat but also, more importantly, to its preservation: 'after many days roasted meat which otherwise would have long have gone to waste is tasty'.[14]

Von den Steinen clearly set out to shock his modern readers by only pointing to the benefits which fire may have given to people in primeval times and by leaving the dangers almost completely aside:

> Here, however, anyone who is used to looking through the glasses of culture will protest. He will miss the horrors which people in prehistoric times have experienced from this tremendous phenomenon and which are little more than the horrors of the scholar whose table lamp may tumble and set fire to his study, his home, and to the entire city with all the valuables it contains. If I, who have tamed the power of fire already start trembling and shivering as soon as the raging element is set loose, if the gigantic play of flames excites *me* already through its impression of fantastic beauty, how must the soul of the poor savage be filled with fright and awe before this mystery.[15]

Even if Von den Steinen has exaggerated somewhat, the gist of his warning deserves to be taken seriously. There is no reason to presume that during the first stage preceding any form of active fire control people could respond to fire only by panicking. We even have to reckon with the possibility that what we today may regard as the natural human reaction to fire is based to a large extent upon our own experiences with fire in

[13] Karl von den Steinen, *Unter den Naturvölkern Zentral-Brasiliens* (Berlin: Dietrich Reimer, 1894), p. 220. One the major attractions of ash is concentrations of salt it contains, of which hoofed animals in particular are very fond.

[14] ibid. p. 221.

[15] ibid.

modern society. The ways in which fire is used have changed in the course of time, and so have the dangers which it poses and the fears it arouses. Fears which may strike us today as natural may be partially a direct consequence of the domestication of fire.

In this respect, the singular form of the word fire may be somewhat misleading. There are natural fires of many kinds, varying with vegetation, season, climate, wind, humidity of air and soil and many other factors. In the era preceding domestication there will have been slowly creeping groundfires as well as blazing forest and grass fires fanned by storm and rapidly destroying all living matter within their reach. Against the latter kind of fire, people were powerless. All they could do was flee. When the fire's raging was over, however, and the charred remains of what had been consumed were still smouldering, it seemed puny and harmless. It radiated an agreeable glow and it let itself be approached without great danger.

Von den Steinen's account shows that humans were not the only living beings to profit from the remains of a fire. Falcons and vultures circling over the ashes and pecking up their prey dead or alive also used the fire to their advantage, and so did all other animals who had come out of curiosity, to warm themselves, or to lick up the ashes. None of these animals did anything, however, to influence the fire, to keep it going, or to direct its course. Only humans learned to control certain caprices of fire and to exploit certain advantages it offered. They learned first to preserve a fire at its original site and later also to transport it to their own camps and dwellings. How did they succeed in this? What enabled them to do so? And why did no single other species achieve the control of fire?

From Passive to Active Use of Fire

The transition from passive to active use of fire has been one of the most important transformations in the history of humanity. This is not to say that it has taken place abruptly, as a 'revolution'. On the contrary, we can best conceive of this breakthrough as a protracted gradual process made up of small steps forward, even when often followed again by a setback. Bands of people who had been fortunate enough to find a fire may sometimes have managed to keep it burning for weeks, months, or even years until it went out as a consequence of rain, neglect, or for whatever other cause; then they would have to live again without fire. Considering how difficult it must have been to keep a fire going through a wet season it is very likely that initially groups of people could only enjoy the possession of fire temporarily and on an uncertain basis.

The question of when and where the transition to active use of fire took place has not yet been settled. Recent finds by a group of archaeologists near Chesowanja in Kenya may lead to the conclusion that humans or human-like primates were already using fire as long as one million four hundred thousand years ago.[16] In close vicinity to hominid bones and stone implements, lumps of baked clay were found which could only have been formed at a much higher temperature than is normally reached in wild bush fires. The find itself and its age are undisputed, but not all experts are convinced that the evidence unequivocally leads to the conclusion that the heat to which the clay has been exposed can only be ascribed to a fire kept by humans and not to some natural source such as a slowly smouldering tree trunk. The oldest deposits which are generally recognised as indicating human use of fire have been found at various sites in Asia and Europe, and date back to about 500,000 years ago.

Although the actual transition to the active use of fire is still shrouded in the clouds of prehistory, attempts at a hypothetical reconstruction need not be entirely unfounded. It is highly unlikely that the transition has occurred as a singular and abrupt event; it is far more plausible that repeatedly groups of human beings had made steps in the direction of a more active use of fire only to lose their embers again after some time. It would be equally misleading to conceive of these early attempts as if they were guided by a conscious ideal of the eventual 'control of fire'; there is more

[16] For the finds near Chesowanja and their interpretation see J. A. J. Gowlett, J. W. K. Harris, D. Walton and B. A. Wood, 'Early archaeological sites, hominid remains and traces of fire from Chesowanja, Kenya', *Nature*, 194 (1981), pp. 125–9; J. A. J. Gowlett, J. W. K. Harris, D. Walton and B. A. Wood, 'Reply to Glynn Isaac', *Nature*, 196 (1982), p. 870; and Glynn Isaac, 'Early hominids and fire at Chesowahja, Kenya', *Nature*, 196 (1982), p. 870. For earlier, less ancient finds in Asia and Europe see Cathérine Perlès, *Préhistoire du feu* (Paris: Masson, 1977), pp. 13–26, and 'Hearth and home in the Old Stone Age', *Natural History*, 90 (1981), pp. 38–41. The hypothesis launched by Raymond Dart in 1947, that charred bones in Makapamja Valley in South Africa point to the use of fire some 1.5 million years ago, is regarded as untenable by most experts. See C. K. Brain, *The Hunters or the Hunted? An Introduction to African Cave Taphonomy* (Chicago: University of Chicago Press, 1981), pp. 54–5; Kenneth Oakley, 'Fire as Palaeolithic tool and weapon', *Proceedings of the Prehistoric Society*, 21 (1955), pp. 36–48, at p. 36; Kenneth Oakley, 'On Man's use of fire, with comments on tool-making and hunting', in S. L. Washburn and Phyllis C. Jay (eds), *Perspectives on Human Evolution* (New York: Holt, Rinehart and Winston, 1961), pp. 176–93 at pp. 176–8; Perlès, *Préhistoire*, pp. 12–14.

reason to assume, as Frederick Zeuner does in his study of the domestication of animals, 'that the beginnings of such processes are largely unintended and not conscious'.[17]

Initially, humans like all other animals will have been forced to flee from blazing forest and grass fires. They will also, again like other animals, have been drawn to the site of extinguishing fires to collect food and to seek warmth and protection for the night. One of the advantages which humans and possibly other primates as well, possessed over all other animals was that they could avail themselves of sticks which enabled them to rummage in the ashes without direct exposure of their skin to the heat. Poking in the ashes in search of food, they could hardly fail to discover in the long run that it was within their power to influence the fire.

When we try to imagine how human beings have started on the long road towards fire control, we may envision them—used as they already were to carrying sticks and hurling stones—poking in a fire and throwing branches into it to see it flare up and to prolong its burning. The ease with which we can evoke such a picture may make us overlook what a remarkable feat it represents. Gathering wood and feeding a fire with it are activities which people perform, not in their own immediate interests, but so to speak 'in the service of fire'.

Especially when people began to collect fuel over larger distances, they devoted part of their energies to maintaining something outside themselves. This does not mean of course that they were in any way acting 'unselfishly', but tending a fire was a rudimentary form of 'detour behaviour', of the pattern of 'deferred gratification' which was later to become a distinguishing characteristic of cattle raising and agriculture, both of which implied the tending of animals and plants in order to reap future benefit from present work.

We may speak of detour behaviour or deferred gratification whenever people attune their conduct to a future goal and restrain those impulses which demand a more immediate satisfaction. Collecting fuel, often with great effort, and building up fuel stocks—in a later stage of the domestication of fire—became almost prototypical forms of such detour behaviour. If the first moves in this direction have necessarily been modest and often abortive, they are nonetheless remarkable.

[17] Frederick E. Zeuner, *A History of Domesticated Animals* (London: Hutchinson, 1963), p. 15.

Remarkable is also the control of fear which must have always played a part in handling fire. Fire is dangerous, and it remains so even when it is looking insidiously harmless. A slow-burning log may suddenly explode in a rain of sparks. Out of softly smouldering ashes, flames may flare up all at once. There are good reasons to be wary of fire in all its guises. If people have gradually learned to control it to a certain extent this is due in no small degree to the fact that they have also learned to control the fears which it arouses in them.

The extended power over fire thus implied an extended power of human beings over themselves, over their own anxieties. In other respects too, the increasing control of fire implied an extension of power. As people managed to make more use of fire their range of life chances broadened. In the course of many generations, they will have learned to roast meat and other food, thereby extending their diet. They will not have been content to collect from the ashes the burned bodies of animals which had fallen the victim to a bush fire, but they will have learned to lay the prey of their own hunt upon the fire. Gradually they will have discovered in what manner and how long various sorts of animals required roasting so that the meat would become tender without being charred.

All these matters may have been learned step by step long before people were capable of transporting fire. This goes to show that the transition from passive to active use of fire has been gradual rather than abrupt. It is very well possible, then, that the two interpretations of the baked clay lumps among the Chesowanja deposits are both correct: that they stem from a 'wild' fire, not laid by humans, but was indeed used and perhaps even tended in some rudimentary fashion by humans.

The use of fire as a weapon, the most formidable extension of human power in prehistoric times, may have originated under similar conditions at a stage when people were not yet able to transport burning matter over long distances but when they were already familiar with the arts of handling clubs and throwing missiles. Given the likelihood that they used sticks to poke in the ashes, it is almost equally probable that they noticed the possibility of making a stick burn on the far end without losing one's grip upon it. Such a burning club formed an extremely dangerous and threatening weapon to sway, to hurl and to beat with. Its invention, no matter how simple and plausible, must have added a tremendous edge to the already impressive fighting potential of a species capable of handling clubs and missiles. One did not need strong physical prowess to inflict serious wounds and heavy pain with this weapon.

It is precisely its destructive potential which made fire so difficult to control; this very fact may also help to explain why the mastery of fire became an exclusively human achievement. This is the subject of the following, admittedly speculative excursion.

The formation of the species monopoly. The capacity to use fire as a torch with which one can deliberately or accidentally kindle new fires opens up enormous possibilities of destruction. Most wild fires are caused by lightning, that is by specific events of short duration which are usually season-bound and are often accompanied by heavy rainfall.[18] Once people had learned to handle fire, however, they could make things burn throughout the year, including dry spells and including the winter. Thus, the frequency and intensity of fires underwent thorough changes, as did the number and kind of experiences with fire which both humans and other animals incur in the course of their lives. From a sporadically occurring natural event fire has increasingly become an instrument of human power, applied at will on a great many occasions.

More fire means more danger of fire. Initially this will have been felt most acutely by members of the human species themselves and by their immediate kin, in so far as the latter were competitors in attempts at exploiting fire. As long as no techniques for transporting fire had been developed, the relatively rare fire-hearths of natural origin will often have been eagerly desired and contested. The fire itself may have been employed as a weapon in the struggle. Those who had settled first near the fire could ward off their assailants with burning sticks lighted in the same fire for which these assailants were coming.

For those groups who had these elementary fire-arms—in the most literal sense—at their disposal it will have been of vital necessity not only to control the fire itself to some degree, but also to keep the use of fire by its own members in check. Only groups capable of developing some caution and restraint with regard to fire, and of instilling these standards of caution and restraint into their young in each generation anew could in the long run maintain fire as a means in the struggle for existence. Internal social control of fire was needed in order to be able to handle fire effectively in the struggle with other groups.

[18] E. V. Komarek, 'Effects of fire on temperate forests and related eco-systems: Southeastern United States', in T. T. Kozlowski and C. E. Ahlgren (eds), *Fire and Ecosystems*, pp. 251–78, at p. 253.

Here a solution may perhaps be found to the problem of how the possession of fire has become a species monopoly of man—a monopoly against which all other animals are defenceless. While humans have grown used to handling fire, all other animals have increasingly come to know it as part of the apparatus of control with which humans dominate and terrorise the animal world and from which every other animal can only seek refuge in flight.[19]

Tertium non datur. In the long run, no middle way has turned out to be possible between the use of fire by one species and the exclusion from such use of all others. We need not rule out the possibility that other primates may have experimented with rudimentary techniques of fire control: throwing wood on a fire to see it burn, setting the end of a stick on fire and swinging it around. If some primates ever started on this route, however, they were driven off it by their more successful hominid competitors.

The idea that other primates may have used fire in a rudimentary fashion in prehistoric times is admittedly speculative but not altogether unfounded. It fits in with the theory of Adriaan Kortlandt that the higher primates living today represent the 'dehumanised' descendants of human-like ancestors who were driven back into the forest by their hominid relatives.[20] Nothing in the physical constitution of chimpanzees would prohibit them from throwing wood on to a fire or lighting sticks in it. One possible disadvantage of chimpanzees and other apes in relation to man could lie in their hairy fur, from which sparks might not be so easily flicked off as from the thinly haired human skin. This possibility is not mentioned by Desmond Morris in *The Naked Ape*,[21] a book which deals at length with the consequences of man's lack of a furred skin but which leaves the issue of fire out of account.[22]

[19] Arthur J. Jelinek, (1967) 'Man's role in the extinction of Pleistocene fauna', in P. S. Martin and H. E. Wright, Jr. (eds), *Extinctions* (New Haven, Yale University Press, 1967), pp. 193–202; S. L. Washburn and C. S. Lancaster, 'The evolution of hunting', in William L. Thomas, Jr. (ed.), *Man's Role in Changing the Face of the Earth* (Chicago: Wenner-Gren Foundation, 1968), pp. 213–29.

[20] Adriaan Kortlandt, *New Perspectives on Ape and Human Evolution* (Amsterdam: Stichting voor Psychobiologie, 1972); Adriaan Kortlandt and M. Kooij, 'Protohominid behaviour in primates', *Symposium of the Zoological Society of London*, 10 (1963), pp. 61–88.

[21] Desmond Morris, *The Naked Ape* (London: Jonathan Cape, 1967).

[22] Adriaan Kortlandt has pointed out to me in conversation that the hair growth of different primates has a different structure. A fatty fur may also offer some protection against burns and against the pain of a superficial contact with fire.

Initially small advantages, in constitution and behaviour, may have been decisive in the success of humans to cultivate and eventually to monopolise the use of fire. The very success of humans may have increased the terror of fire and the incapacity of handling it among competing species. This, then, might account for the remarkable fact that the control of fire is an exclusively human acquisition. If this hypothesis were correct, the fire monopoly, like later intra-specific political and economic monopolies, would be the outcome of a struggle of elimination and as such, once the struggle had been fought, so firmly established as to seem self-evident.[23]

Preconditions for the Active Use of Fire

Half a million years ago, groups of people at widely remote places in Eurasia were already capable of tending a fire. Archaeologists have found remains of fires human kept by humans in sites at Zhoukoudiem or Chokoudien[24] near Peking, at Vertesszollos in Hungary, and at Arago in Southern France. It is not known whether the groups who lived at these sites had arrived at a more active use of fire independently of each other or through borrowing and imitation.[25]

The most solid and complete evidence collected thus far about the early use of fire stems from the famous cave at Zhoukoudiem where excavations have been made since the early 1920s. Traces have been found of an almost uninterrupted occupation of this limestone cave by humans (*Homo erectus*, a closely related predecessor of *Homo sapiens*), over a period stretching from approximately 460,000 to 230,000 years ago, when the roof collapsed.[26] Most of this period, the inhabitants had been tending fires in the cave. Four layers of ashes have been found among the deposits, the thickest layer being almost six metres deep in places.[27]

[23] It can be instructive to consider the analogy between inter- and intra-species monopolies, even though such an analogy is obviously limited. On the formation of political and economic monopolies in human society, see Elias, *On the Process*, pp. 301–11.

[24] These are the romanisations used by Goudsblom; alternatives are Zhoukoudian and Choukoutien.—eds.

[25] See above, note 16.

[26] The length of Peking Man's occupation of these caves has, however, more recently been very much disputed, as Goudsblom noted in his later book *Fire and Civilization*, p. 17.—eds.

[27] For a brief survey of the finds, see Chia Lan-Po, *The Cave Home of Peking Man* (Peking: Foreign Languages Press, 1975); Wu Rukang and Lin Shenlong, 'Pekìng Man', *Scientific American* (June 1983), pp. 78–86. For a critical comment upon some earlier interpretations,

Two facts in particular are worth noting. First is, that the fires were kept at a place where they could not possibly have originated spontaneously. They must have been lit deliberately, in all likelihood by means of a burning torch—perhaps a bundle of twigs as Rukang and Shenlong suggest—which had been carried away from a site where a natural fire had burned.[28] This transport of fire amounted to a decisive change in the relationships between humans and fire. Evidently, people had extended their control of fire in such a way that in order to benefit from it they no longer had to stay at the spot where a fire had happened to occur naturally but, instead, they could now take it with them to a place which suited them best. The second remarkable fact is the truly staggering continuity of over two hundred thousand years during which fires were kept at one and the same spot.[29]

Both facts are obviously connected. Once people had learned to harness fire, they derived certain advantages from it, not the least of these being that they could conquer and defend caves and make them into permanent dwelling places. While, on the one hand, the possession of fire made it easier for them to settle in caves, the occupation of a cave, in turn, greatly facilitated the protection of fire against rain and wind and against raids. Living in the relative security of such a fire-protected abode, people could hand down to their offspring the knowledge and skills needed to maintain their privileged position. This is an excellent example of the vicious circles in which human groups have found themselves again and again in the course of their development, and which sometimes worked to their detriment, but sometimes, as in this case, also to their advantage.[30]

The development of the control of fire abounds with such spiral effects. When we ask ourselves what sorts of conditions have made this development possible, it is not sufficient to mention one or two separate 'factors' to which a decisive influence might be ascribed. We are faced, rather, with

see Lewis R. Binford, *Bones: Ancient Men and Modern Myths* (New York: Academic Press, 1981), pp. 283–92.

[28] Rukang and Shenlong, 'Peking Man', p. 85.

[29] As noted above, Goudsblom later doubted this 'remarkable fact'.—eds.

[30] The concept of a vicious circle is used here without the connotation of 'bad' and 'perverse' which are etymologically associated with the word vicious. For discussions of vicious circles in social processes see Gunnar Myrdal, *Value in Social Theory* (London: Routledge & Kegan Paul, 1958), pp. 188–92; Raymond Boudon, *The Logic of Social Action* (London: Routledge & Kegan Paul, 1981); Norbert Elias, 'Problems of involvement and detachment' [orig. 1956], in *Involvement and Detachment* (Dublin: UCD Press, 2007 [Collected Works, vol. 8]), pp. 68–104.

a set of mutually reinforcing conditions, some of which may be regarded as 'preconditions' enabling people to learn to master fire, and others as 'post-conditions' or 'functions'—consequences which made it worthwhile for human beings actually to realise and exploit the available possibilities, and which thus, in turn, positively influenced the 'preconditions' that were already given.

In this chapter, I shall focus on the 'preconditions', the 'functions' will be considered in a following chapter.[31] The development of the 'preconditions' cannot be seen in isolation from the general evolution of the human species. There is a set of closely connected features which have become distinguishing characteristics of human kind, including an erect posture, flexible hands, a highly differentiated brain and nervous system, and a long learning period spent in close association with other humans. It is this set of mutually related traits, developed in the course of human evolution, which has enabled human beings to control fire. Together they enter into the entire set of activities which is involved in even the simplest active use of fire: finding an existing fire, obtaining a stick or a bundle of twigs to capture some of it, carrying the captured piece of fire to a safe and dry place, and tending it by supplying it with air and fuel.

The significance of an upright position for the domestication of fire seems fairly obvious. It is hard to imagine how people could have picked up and transported fire if they had not been able to walk on two legs and if, instead of carrying fire in their hands, they had had to carry it in their mouths, in the way in which, according to certain myths, the robin and the wren had once carried fire in their beaks.[32] Hands, free to carry and

[31] See Goudsblom, *Fire and Civilization*, pp. 37–8.—eds.

[32] For myths on the wren and robin as 'fire birds' see Edward A. Armstrong, *The Folklore of Birds* (London: Collins, 1958), pp. 175–79. Michael Allaby writes in his book *Animal Artisans* (New York: Knopf, 1982) that the black kite (*Milvus migrans*) picks up smouldering twigs from spontaneous fires and drops these upon dry grass in order to spread the fire and to catch fleeing animals. In recent issues of the *Zoological Record,* many aspects of the life of the black kite have been described, but no mention is made of the behaviour mentioned by Allaby, nor is it reported in the major zoological reference works. The Dutch expert on birds of prey, Professor K. M. Voous, drew my attention to the possibility that black kites hunting a savannah fire ('a true feast for birds of prey') for large insects which have been sent up by the hot air of a fire may sometimes by mistake grasp a burning twig in their claws and fly away with it. They will then have to drop it again soon. Allaby's statement may therefore be grounded upon observations which are correct, but his interpretation is questionable. The same holds for the claim of the biologist, Maurice Burton, in *Phoenix Re-born* (London: Hutchinson, 1959), that rooks and other birds may occasionally carry burning embers of

sufficiently flexible for performing subtle and complicated operations, have become indispensable in the art of 'handling' fire: first for collecting and transporting fire, later for igniting it with the aid of wooden or stone implements.

In contrast to the pattern of nest building by birds, or the inborn capacity of certain animals to attack stronger and bigger enemies precisely at their weakest spot, the art of dealing with fire does not spring from any specific genetic code. The ways to handle fire—this hazardous element—have to be altogether learned. The initial learning process in the Old Stone Age can only have been gradual and slow. People in the immediate surroundings of a fire may have busied themselves with burning sticks; they may thus have kindled new fires, first inadvertently and later on purpose, and in doing so they may eventually have discovered means of transporting fire over larger distances, of 'stealing' it as the Prometheus myth and many other myths suggest. Whereas most myths refer to the heroic feat of a single individual who 'gave fire to man', it seems more realistic to think of people in groups who, in the course of many generations, have learned collectively to deal with fire. Vying with each other in daring and skill, encouraging and warning each other as in a communal hunt, teasing and chasing each other and, especially, watching each other all the time, they will have learned little by little from one another's experiences.

Perhaps other primates have also ventured upon this dangerous social pastime. Eventually only humans, however, made the decisive transition from passive to active use of fire. What has enabled them to do so is the combination of specific physical traits and the disposition to learn from others. The latter has been, and continues to be, essential. Even in the absence of experiments in this matter we may safely assume that descendants of contemporary *Homo sapiens*, equipped with all the genetic material of modern humans, would be very unlikely to acquire the art of handling fire if, from the day of their birth, they had never encountered any examples of it.

The example set by others is indispensable. In this respect the control of fire is a typical element of culture—'learned, shared and transmitted'

some sort or another to their nests, thereby sometimes causing chimney and roof fires. Burton presents fascinating evidence, including photographs, of the predilection of some birds for fire, which clearly shows that the natural reaction of animals to fire is not just fear. His evidence for birds actually and deliberately carrying fire in their beaks remains inconclusive, however.

according to Tylor's classical definition.[33] The domestication of fire has been a long-term process involving numerous generations. In the course of this process people at times must have tried something new but, most of the time, they will have sufficed by following one another's example. What had been initiated successfully by some was either imitated by others, or lost to be rediscovered later or perhaps never again.

The mechanism of cultural transmittance may help to explain the continuity with which fires were kept for many centuries in the cave at Zhoukoudiem. The art of tending a fire was handed down here from generation to generation. This involved the imparting, not only of technical skills, but also of certain insights and emotional attitudes—all of which could only be developed by learning from others, in a social context.

In her survey of the use of fire in prehistoric times, Cathérine Perlès notes that 'the discovery of the utilisation of fire presupposes a mental and not a technical progress'.[34] The emphasis on the mental aspect is well taken; but there is something unsatisfactory in the fact that she contrasts the 'mental' and the 'technical' and does not explicitly mention the 'social' component in both. Learning to control fire is a social-psychological process, in which technical, intellectual and emotional aspects are closely interwoven.

The control of fire implies a combination of knowledge, courage and skill. In dealing with fire it is of vital importance that people can command an appropriate dosage of daring and caution, that they do neither immediately panic nor grow reckless and forget how dangerous fire is. A certain measure of self-control is part and parcel of technical control—this applies to tending a fire just as much as to any other human skill. The emotional security with which people can approach fire is furthered in turn by the knowledge and dexterity which they have acquired and on the basis of which they can be confident that they are indeed able to master the fire, to let it flare up, smoulder, or go out at will.

The very fact that fire is so dangerous makes incessant wariness and restraint imperative in dealing with it. Without necessarily subscribing to the view that men have first learned to control fire by suppressing their spontaneous urge to quench it by urinating, we may concur with Freud that the mastery of fire involves a certain watchfulness and, therefore, a

[33] Edward Burnett Tylor, *Primitive Culture* (London: John Murray, 1871), p. 1.
[34] Perlès, *Préhistoire*, p. 30.

reduction of spontaneity.[35] It is as fascinating to watch a fire as it is tempting to play with it; but if one is not careful one may easily inflict terrible wounds upon oneself and others. Learning to handle fire is therefore also a training in cautiousness and consideration. If people generally acquire these qualities without any great effort, this is due to the fact that they grow up in societies which have already attained a certain measure of fire control so that people at an early age can become familiar with the appropriate standards of knowledge and skill, of calm and courage, with regard to fire.

Reflections upon the social-psychological preconditions of fire control inevitably have a strongly interpretative, if not speculative, character. There is no possibility to put them to an experimental test, especially when they concern the very first stages of the development of humankind. The least that can be said for the interpretation given here is that it is plausible and makes sense. It is not at odds with any known facts and it does throw some light upon the general question of how the control of fire has come to be an exclusive and universal human acquisition. The ability to handle fire is shown to be preconditioned by a set of mutually reinforcing traits which have resulted from a process of biological evolution and social-cultural development.

It is very likely that the formation and transmission of culture within human groups has been stimulated by another, no less important social mechanism—the competition between groups. The fact that of all subspecies of hominids only *Homo sapiens sapiens* have survived to the present day intimates that the struggle for existence must have been hard. In this struggle, in which even seemingly small advantages may have been of decisive importance,[36] the capacity to handle fire more effectively than others must have made a considerable difference—not only for the relations between human beings and other animals but for the relations of human groups among each other as well. Once certain groups had gained an advance in the control of fire, neighbouring groups could not afford to lag behind. It stands to reason that those groups who did not succeed in

[35] Sigmund Freud, *Civilization and Its Discontents* (New York: W. W. Norton, 1962).
[36] See above, p. 223. See also Washburn and Lancaster, 'Evolution of hunting', p. 221.

taking over the art of using fire were doomed to perish, by way of subjection or extinction, or perhaps of flight and regression. In the long run, the co-existence of groups of people with or without fire has turned out to be impossible. Thus, fire has become an exclusive *and* a universal human attribute.

Discussion of Goudsblom's Paper

Transcript

Elwert (in the chair): The provocation of the paper lies in the central thesis that hominisation centres around the social organisation of fire management and fire control. And I think that's a very good thesis …

Wallerstein: I have a question, less about fire control than about the consequence. And I found the most fascinating suggestion not that fire control was an essential element in the civilising process but that as a consequence of it there was the de-humanisation of some hominids. That is to say, in effect, you're telling us that there came to be a new social definition of what came within the realm of humanity. I'd like to know more about what that means operationally. You mentioned retreating into the forest, which I can see as being defined or having something to do with de-humanisation. What else would you think of a being included in the process of de-humanisation of hominids who were somehow human before? Did they lose characteristics which we associate with human development? Did they perhaps have more language than they had later? Could you elaborate that, in any case?

Transcript (✉)
Berlin, Germany
e-mail: Artur.Bogner@web.de

A. Bogner, S. Mennell (eds.), *Civilisations, Civilising Processes and Modernity – A Debate*, Palgrave Studies on Norbert Elias,
https://doi.org/10.1007/978-3-030-80379-7_15

Goudsblom: Well, that thesis under the name of de-humanisation has been formulated, as I said before, first by my Amsterdam colleague Adriaan Kortlandt.[1] And, as I did my study on fire, it suddenly struck me—he never mentions fire at all.

It suddenly struck me that the two things I had already formed in my own mind about how the development of fire control might have taken place—that it somehow fitted into his theory about de-humanisation, of which I cannot tell you the biological details. I mean, there is anatomy involved and physiognomy. But what I do know is—what I've found most easy to grasp is the behavioural aspects that he mentioned. And one of these behavioural aspects is for example the capacity to throw and to aim. And it was generally supposed by all zoologists that chimpanzees were incapable of aiming and throwing. But then, in a zoo, someone decided that it might be a good idea to just experiment and try if they could not be taught the art. And in a surprisingly short time, not only did they learn to handle missiles, but also they developed a very accurate aiming capacity. Whereas other apes who were trained just didn't learn. I mean it's a very complicated ... It's one of those things about which we don't stop to think how complicated it is. The act of picking up a stone, taking aim, throwing it, and hitting the target. And chimpanzees turned out to be capable of learning this very rapidly. And that was one of the things that gave Kortlandt the idea that, phylogenetically, this must have been something that was present, the potential was there. But it had been lost. And he also [found] there are also other capacities that have been discovered that chimpanzees possess, that they have not been observed to realise in their natural surroundings or in zoos until they were taught to realise them. That gave Kortlandt the idea that if this is given phylogenetically, apparently at one time they had developed these capacities. They had a survival value for them but their habitat has changed, their habits have changed, and the skills have become completely obsolete, and they have not any longer been taught from one generation to the next. They have no longer been transmitted, but a genetic capacity has not been lost.

Wallerstein: And the shift of habitat is essentially from the plains to the forest.

[1] Adriaan Kortlandt, *New Perspectives on Ape and Human Evolution* (Amsterdam: Stichting voor Psychobiologie, 1972); Adriaan Kortlandt, and M. Kooij, 'Protohominid behaviour in primates', *Symposium of the Zoological Society of London*, 10 (1963), pp. 61–88.

Goudsblom: Yes, because in a plain, throwing is a very effective way of hunting. Whereas in forest hunting, certainly in a dense rain forest, you can't throw. It's completely irrelevant. So that is one of the arguments. Was this all that was implied in your question?

Wallerstein: No ... his evidence is all for the phylogenetic potential?

Goudsblom: Yes, indeed ...

Wallerstein: There's no archaeological evidence in addition?

Goudsblom: There is archaeological [evidence]. Well, one wouldn't call it archaeological anymore, I think. But there is indeed—there have been many bones and pieces of skeleton have been found that point to a common origin. But that is a part of the story that I consider myself not sufficiently competent in. But what I have then here given is a sort of—well, I use the word palaeosociology jokingly—but it is a sort of attempt at a sociological reconstruction.

Bertelli: Only a marginal observation. The aborigines of Australia use fire in order to capture their food. They burn the bush for hunting. Nevertheless, they don't control the fire. Neither do they take care of it ... They have not developed themselves in the sense of a process of civilisation as we understand it. In fact, at least in this case, the discovery of the use of fire by itself is not in my impression sufficient to start a process of civilisation.

Goudsblom: May I quote from the next chapter that I have written? Because I have also gone into the problem of the fire control in later times. Here is a quotation from an explorer in 1827, Scott Nind, no, not an explorer, who worked as a health officer in Southeast Australia in 1827 and reported on the life of the aborigines. 'At this season', at the end of the dry season,

> they procure the greatest abundance of game ... by setting fire to the underwood and grass which, being dry, is rapidly burnt. ... With a kind of torch made of the dry leaves of the grass tree they set fire to the sides of the cover by which the game is enclosed. ... The hunters concealed stand in the paths most frequented by the animals, and with facility spear them as they pass by. On these occasions vast numbers of animals are destroyed. The violence of the fire is frequently very great and extends over many miles of country; but this is generally guarded against by burning it in consecutive portions.[2]

[2] Quoted in Goudsblom, *Fire and Civilisation* (London: Allen Lane, 1992), p. 32, from Sylvia J. Hallam, *Fire and Hearth: A Study of Aboriginal Usage and European Usurpation in*

So they are very careful. And there are other quotations that I have found pertaining also to the Australian aborigines. Another man, J. L. Stokes, around 1840—who also travelled on the *Beagle* and wrote his own diary—says:

> On our way we met natives engaged in burning the bush, which they do in sections every year. The dexterity with which they manage so proverbially a dangerous agent as fire is indeed astonishing. Those to whom this duty is especially entrusted and who guide or stop the running flame, are armed with large green boughs with which, if it moves in a wrong direction, they beat it out. … I can conceive no finer subject for a picture than a party of these swarthy beings engaged in kindling, moderating. and directing the destructive element, which under their care, seems almost to change its nature, acquiring, as it were, complete docility, instead of the ungovernable fury we are accustomed to ascribe to it.[3]

Bertelli: But they do this also today. I was in Australia. I know how they burn the bush, but this doesn't matter because they didn't develop anything.

Goudsblom: Then I think I missed [your question]

Wallerstein: I think the question is, once you get the fire, do you get any further?

Goudsblom: I misunderstood your question. I shall return to it.

Mennell: Well maybe be the opportunity has passed. Maybe I'm jumping to tomorrow's agenda for comparing theories. But I wasn't at all surprised that Immanuel Wallerstein was so interested in this proposal about the possible process of de-humanisation, because it strikes me there is some kind of theoretical homology there with the kind of argument that crops up time and again in dependency theory. For example, Andre Gunder Frank's famous remark that the underdeveloped world is underdeveloped because the capitalist world has underdeveloped it;[4] or, to take an example nearer home, the discussion in your first volume of *The Modern World System*, Professor Wallerstein, where you talk about the relationship between eastern and western Europe from about the fourteenth–fifteenth

South-western Australia (Canberra: Australian Institute of Aboriginal Studies, 1975), pp. 32–3.

[3] John Lort Stokes, *Discoveries in Australia*, 2 vols (London: T. & W. Boone, 1846), vol. 2.

[4] Andre Gunder Frank, *Capitalism and Underdevelopment in Latin America* (New York: Monthly Review Press, 1967), p. xi.

century.[5] And it certainly seems to me that these are excellent examples of the kind of compelling proof that there is a theoretical homology very obviously between what Professor Goudsblom has been saying, the kind of conceptual scheme that Professor Elias has been developing and some of the ideas which Professor Wallerstein uses, too.

Brands: May I just ask about the word 'domestication'? I don't know [quite how] to put my question about, on the one hand, the increased control of fire but, on the other hand, the surprise now at the increase in the power of the uncontrolled fire. How do they fit together into your scheme? I mean, every year we still do have a major surprise by fire that, for specific reasons perhaps, cannot be controlled for a long time, for a month for instance. How do they fit into that? And second, but that's quite a different question, I was thinking about burning human beings. That's a special aspect of course: [burning them] not on purpose, and later on, on purpose. Will that come up in some chapters as well?

Kilminster: Your example of the vicious circle is that the people living in these caves—on the one hand you say the possession of fire made it easier for them settling in caves, and the occupation of the cave in turn greatly facilitated the protection of fire against rain and wind and against raids. Now, you then go on to say that that's a vicious circle and in the footnote you say, the term vicious by that I don't mean any connotations of bad or anything like that. And what struck me about that was what you've got there is something like a circular process of reinforcement or some such term one will have to use, feedback … I think substantively what you say there is correct, but I personally would question using the concept of a 'vicious circle' to describe it. Because the consequence you describe there seems to be quite benign. That's one point.

Then, right at the end of your paper, where you start to talk about the sociological implications of this, there is a penultimate sentence that I'd like you to clarify for me, please. You say, 'It stands to reason that those groups who did not succeed in taking over the art of using fire were doomed to perish, by way of subjection or extinction or perhaps of flight and regression'. Do you mean 'regression' or do you possibly mean 'repression'? And if you mean 'regression', what were you referring to? Do you mean that the groups would return to an earlier stage of development,

[5] That is, the suggestion that commercial trade with western Europe caused eastern European agriculture to 're-feudalise'; see Immanuel Wallerstein, *The Modern World-system*, vol. I (New York: Academic Press, 1974), pp. 95–6.—eds.

or what do you mean? That sentence strikes me as being quite a crucial one and I may have misunderstood it.

And my last point again is as specific as that, and may sound at first sight formalistic. But I don't think it is. If it is, may I be forgiven? You say in the first paragraph, 'The control of fire is always part of a system of social control and self-control'. Now, I would go along with that. Now, do you mean fire or control?

Goudsblom: Control.

Kilminster: The development of the control of fire 'may be regarded as a "civilising process" in Norbert Elias's sense of that concept'. Now, I was wondering if it's not so much the control of fire that is a civilising process, but rather that the control of fire is part of a civilising process.

Goudsblom: That's much better. That is more correct.

Kilminster: So can you comment on those other two questions?

Goudsblom: May I speak now because otherwise it will really become too much? I shall start with Professor Bertelli. I'm sorry I misunderstood your point and I thought you meant to say that these aborigines were not really civilised in handling fire, but you said they 'they have not developed themselves in the sense of a process of civilisation as we understand it'.

Bertelli: In this case, doesn't start the process of civilisation. In fact, it is not sufficient in itself.

Goudsblom: That is then, I'm afraid we are again using the concept of civilising process and civilisation in a different way. What I tried to do here is to use the word 'civilising process' and the word 'civilisation' in the same way that anthropologists have been using the word 'culture' and I prefer the word 'civil*isation*' and 'civilising *process*' because 'culture' is a static and 'civil*isation*' and 'civilising *process*' are dynamic concepts. And what I want to say is that everything that human beings learn from others, and transmit again to others in the way of codes of conduct can be regarded as a civilising process. So every individual human being goes through a civilising process. The member of an aboriginal tribe in Australia had to go through a civilising process in order to learn to handle fire in this very skilful manner—as it has been often described, and as you apparently have still even been able to observe it. And the fact that the aborigines have reached that high standard of fire control can only be explained by saying that they, as a society, have gone through a process of learning—no matter what you call it—and I would say that this process of learning to handle fire, which involves a great deal of self-restraint and also a great deal of social coordination; among the Australian aborigines this is very clear because it's a very delicate thing. I mean, once you set fire to a land to which you have claimed property rights, you have to see to it that the fire does not extend to the land owned by another group. And it can easily, of course, go too far and then it can become the source of a very violent conflict—and a conflict can, again, be fought out with fire. I got into that and, in a later chapter, I try to explain that one of the functions of the use of fire is in land clearance and keeping the land open for game and hunting, as the aborigines did. But that implies certain precautions and certain social regulations that have to be observed. So there is a disciplining, a civilising, compelling civilising force which is an integral part of this whole culture, economy. 'Fire economy' it has sometimes been called, 'fire culture' it has been called. It dawned upon me only after your second intervention, that your question is, if I understood it correctly, why, if they were already so advanced in their control of fire, didn't they develop any further?

Bertelli: Yes, yes.

Goudsblom: But that, of course, pertains to the whole development. If that's the [case], what does one consider an explanation? A sociological explanation? That's really the question behind your question, the more general problem, pointing at preconditions. That is what I set out as my major task, exploring the simple fact that we can now handle fire, and I shall come to the dangers that Maarten Brands pointed to in a moment. Of course, it's still not free of danger, but usually we manage very well and most of us will not have been the victim of a fire in the last ten years. I say 'most of us'. What I wanted to ask myself is, what preconditions had to be met for human beings to get that far? And then I immediately found that once I started looking into the preconditions I could not ignore what I called the 'postconditions'. It's not a very nice word and maybe we can do away with it. But the line of argument, I think, is clear. What it does not at once convey is the chronological unfolding of events. I mean, that would require a much more detailed [account], and I shall try to go into it. I shall try to show in the following chapters how fire was a precondition for the next great transformation in human life, the domestication of plants and animals. I hold really, as a very strong statement, that domestication, agrarianisation, would have been impossible without the control of fire. There you have a sequential order in the development of humankind that is of a very strong kind. But one *cannot* say that once human beings had acquired fire, were able to control fire, agriculture had to follow necessarily. That is, that's another kind of question because there again once you come to the great transition, when human groups started to practise agriculture, you have to ask again, what were the preconditions and what were, so to speak, the postconditions? What made it rewarding and perhaps compelling for them at that particular stage to engage into this rather unpleasant work involved in agriculture? And the Australian aborigines never reached that stage. … So I would say your question is certainly legitimate, but one should not allow oneself to say that, in dealing with the very first stage and answering the questions that the very first stage raises, one should already face the problem of why, in a much later stage, the people on the Eurasian continent engaged in agriculture. And why did it happen around 10,000 years ago? And why did the Australian aborigines not do it? I think that's a different, perfectly legitimate and interesting question, but it does not belong in the preliminaries, and I don't think the fact that the Australians did not enter the stage of agriculture in any way undermines the argument involved in this paper. That would be my answer to your question.

[To] Maarten Brands: the word 'domestication' is in a way an anachronism when you apply it to caves. But I think I can be forgiven for that. And bringing fire into a cave I think may be called 'domestication' by analogy. When you apply the word 'domestication' of plants and animals to the very first stage, it's also slightly an anachronism. And the analogy does not fully apply, because I think it can be said with great precision again, that the domestication of plants and animals implied really the change of the balance of power, certainly between the humans and animals and, perhaps one might even say, between humans and plants. Whereas it is actually wrong, I think to speak about a balance of power between humans and fire.

But you say the increase of the control of fire has also sometimes faced us with increasing danger from fire. I think that is certainly a very correct observation and one with which I also shall deal in later chapters. It does not violate the thesis of this paper at all. It does relate also to the discussion we had in the previous section, that extending the human control over physical forces, such as fire, does not mean that these physical forces change in their actual nature. And bringing fire into cities for example—and certainly bringing it into pre-industrial cities largely constructed out of wood—is almost inviting fire. So you find the history of mediaeval cities is a history again and again of great fires, as you know much better than I do. It's also the history of an increasing effort on the part of city governments to counteract those fires through regulations. For antiquity, I am sure you will find, if you really go into it but Keith Hopkins—I've already asked him the question—agrees that so far perhaps the sources are insufficient to really give us a good overview of fires in cities in antiquity. But it would run against all likelihood if they had been absent, and if only the few that have become famous like the fire in Rome of which Nero—wrongly I think—has been accused of being the instigator. But that certainly was not the only one.

And then also the question of burning human beings. That is a very intriguing aspect that indeed also reflects upon the many uses of fire. Once this force, this physical force, is under human control, they can not only burn plants, wild vegetation. And they can not only ward off animals with fire. They can also use it as a weapon against each other. And they can use it to torment their victims, to test their victims. They can use it in many ways. Incidentally, that is an aspect that I have not yet worked out and I hesitate whether I should do it. But I have the intention of not going into any theoretical debates and issues in this book. But, of course, it's a marvellous entrance, a marvellous opening in a debate with the Marxist theory, that means of production are so important. Here you have something that can be used both as a means of production and a means of destruction. And it's one of the central elements already implicit in this paper, that fire as a means of destruction has been of vital importance in the development of human societies.

Then, Richard Kilminster, I agree that the concept of 'vicious circle' is not very fortunate in this context. The word 'beneficious circle' I would also not use. I would want a neutral term that would encompass both. Because what I really want to convey is that it is one of the basic mechanisms of social development: again and again, human societies enter into a stage where mechanisms of this kind are at work, which—as facilitating or de-facilitating forces—drive a human figuration in a certain direction. And it is one of the keys to the whole problem of the unplanned nature of long-term social processes. So you may say it's a merely conceptual problem, and it is, but since it concerns a concept that is so central to the theory, I think it's a conceptual problem worth thinking about.

There was one other point you made regarding my remark that 'It stands to reason that those groups who did not succeed in taking over the art of using fire were doomed to perish by way of subjection or extinction or perhaps of flight and regression'. This refers back to things I've said before. At a time when human bands were still very small, it's unlikely that a whole group could have subjected itself to a sort of a slave existence, but extinction is certainly a possibility and we know that in pre-history many human groups and even varieties of the human species have died out.

Flight and regression? … we are dealing not with centuries and not with millennia, but with tens and hundreds of millennia. It may very well be possible that groups without fire gradually receded from centres where other human groups were living who already had fire. So that's what I mean by 'flight'. And 'regression' refers to the point that was raised by Immanuel Wallerstein.

Koenigsberger: I want to come back to the chimpanzees. I'm sorry. This is not frivolous but may sound so. I really fail to see the point of the chimpanzees. If you told me that the chimpanzees' mothers, having learned to throw balls or whatever, then teach their young to do that, I would be slightly more impressed. I would be impressed, but not very much, because, after all, cats teach their kittens to use claws. I would be impressed if you then led these zoo chimpanzees out into the jungle again, and they used certain of their newly learned throwing abilities in order to establish themselves over other chimpanzee groups. And honestly, I cannot see why we should, in order that it is most unlikely. After all, chimpanzees can also learn to ride a bicycle and I've seen bears riding motorcycles—that's even more complex. It means nothing. What I'm driving at is this: why should we deign to use a purely speculative regression from *Homo sapiens* or *Homo erectus* into some animal status because of fire? When we have Professor Elias's model, or mechanism, by which we know that if you have one group it is likely to clash with another group of the same species, especially if they're human and if they have the advantage of fire, having caught that first, the likelihood, if we have to speculate this, that they will subject them, make them slaves, make them servants, exterminate them or simply drive them away, until these people have learned the same thing. Applying Occam's Razor, why do we need another theory from the one which we have been discussing for two days very happily?

Goudsblom: May I answer this one very quickly, because I think the experiment you suggested is marvellous but impracticable. About the chimpanzees—I put it into the paper very explicitly as a digression, and I know that biologists are interested, I might even say fascinated, by this sort of reasoning. This whole field is a very lively area indeed, filled with speculation. And this was my little contribution to that discussion. But it's not the focus of the paper.

Koenigsberger: I'm glad. Thank you.

Cavalli: I just would like to make you a suggestion that there's an intriguing fact that fire is sometimes used as a political symbol. And generally, it is used as a political symbol by right-wing groups and rightist parties. And this is probably because of the fact that fire, you know, has an intrinsic, symbolic ambiguity, which you pointed out—because it means both energy and destruction. And this very ambiguity that the concept of fire evokes in our culture is probably one of the reasons why we are so hesitant now—and maybe we wouldn't be hesitant 30 or 40 or 50 years ago—to speak about 'progress'. I mean, the word 'progress' has been pronounced over these days [of the conference] with a lot of hesitation. Why are we so hesitant to speak about progress? And I think this has to do with fire. We are hesitant to speak about progress because fire in our civilisation at this stage of this development is the way that humanity has acquired the capacity of destroying altogether the human environment. I mean this is the limit, [the extreme] expansion of the ambiguity of the concept of fire. I think it has something to do with our hesitance in pronouncing the word progress in this context. …

Korte: I think it is very necessary to start the study of long-term processes with these first steps. But, and this is my bias, I'm more interested in the end of these long processes, with the fate of the human beings today. I know that you have several chapters coming, but I would be very thankful if you can give us a little outlook about what that means for societies today, at least. Because the connection, or the interdependence, between psychogenetic development and sociogenetic development is very well worked out here in your paper again. But what does that mean for the analysis of today's societies and what does that mean, what can one say, what can one learn, to make the people better handle their fate? This is not an argument against that paper. I want to make it very clear. I think every time in discussing these long processes one should also have a look to the end. …

De Swaan (in the chair): Well, the chairman should use your paper [to be] provocative: it contains the claim that there is a universal and exclusive human characteristic, which is about the boldest claim one can make in an intellectual environment. And yet I've been struck by the fact that there have been no contradictions except the smallest factual detail, a minor matter of spelling of your statements. In the way I have the idea that we are party to a paradigmatic exercise, even a rather bold piece of sociological conquest of new fields. And do you know why there is so little opposition? Could it be that your points are absent?

Goudsblom: I shall answer in the order in which the questions were raised. I'm not immediately able to order them in any other fashion. Maybe that shows my inclination to good thinking in sequential order.

What Alessandro Cavalli said about the ambiguity of fire is indeed, I think, very well taken. I've tried to show that it's one of the threads running through the paper. I think you are also right in saying that we are so hesitant about progress because you might say, 'Indeed, the human capacity to control fire has progressed'. But the moment you say that you have to realise that this also implies a progression in the danger of fire and in, of course, the danger of uncontrolled fires—as we know in the oil refineries and so on. And the danger that fire may be used on a tremendous scale as a weapon. So, if the word 'progress' is used to apply to such technological changes—and it is not, that is what I want to say among other things—any so-called technological development is at the same time a social-psychological development. If this development with regard to fire has occurred, we sometimes hesitate whether we should congratulate ourselves so strongly with this development. So, I agree with you. The question why fire would be used especially as a symbol by right-wing political movements, if you will pardon me I shall not go into that. I have not given that any thought yet. I don't know. I'll just make a note of it.

Korte: It is also used by the Olympic movement so. …

Goudsblom: Then Hermann Korte's question. What does it mean for contemporary society? I wish I could give you a ready answer to that one. I know that, for one thing, it is important anyhow to see that in the infrastructure of our societies there are certain elements that are sometimes overlooked that are not regarded any more as belonging to that cultural infrastructure. And it's important to bring them into the open again. So, but that's for you. You want more. There are other aspects to it, as well. Abram de Swaan has said it's a paradigmatic exercise. Perhaps it is. It certainly can be used as a paradigmatic case study in human development over a long period. And it shows, as Cavalli has observed, the ambivalence of technological and social development. As I continue, when you enter into the agrarian stage 10,000 years ago, you enter into the stage, first, of slash-and-burn agriculture. And there, you can see again, how double-edged, for my point of view, this may look like present-day society, but you certainly didn't mean that. But you can see how double-edged the use of fire was because, as everyone knows, slash-and-burn agriculture yields very high harvests for one or two years and then the fields are exhausted. And the forest is gone. And the agriculturalists have to move on. And, as long as the population in a certain area is not too dense, after a generation, they may return to the original site and there a new forest may have regrown. So they may start anew. But it's in the nature of agricultural societies, that's again one of your vicious circles, if you want, that they do not only yield a produce in crop, but they also yield a produce in human beings and an increase in population. So, eventually, it's a self-destructive system and people then have to ... But then, new solutions have been found, as we all know, with ploughing and other more intensive means of agriculture.

Evers: Sorry, but there really is one point. I thought perhaps you might be interested to point out that in a civilising process, there's one point telling when fire suddenly becomes hidden. People put it into a fireplace and then make it an oven and then it sort of disappears as the centre of the household. I don't know what the significance of this is, very late time ...

Goudsblom: It's a very late time. If I may take in my stride the next 10,000 years in a few sentences: ... the process of further domestication of fire is accompanied by further differentiation. When you have agricultural towns, there is the blacksmith, there is the baker, there is the potter, and there are all the private households, all having their fires. Maybe there are also still communal, big communal fires. But there is a differentiation, both—you can say both—according to structure and function, there's a differentiation in fire. And with that goes the need for a greater coordination in managing all these different fires, which are all potential sources of conflagrations. And, in our societies, indeed, fire has been domesticated if, one could say in a literal sense, very much so, as you say it has been. Today, children—in England, as Eric Dunning remarked in private conversation, it's different—but on the continent, children may grow up and see a fire only on festive occasions or if their parents smoke. It's something that becomes more and more remote from daily affairs and you only have the differentiated forms of fire in furnaces and in churches, and, I mean, a whole variety of forms. But I would really rather not comment at the moment upon what, in contemporary society, are the many uses of fire.

Korte: Or the many menaces there are of fire. Maybe nuclear energy is the fire of our age.

Goudsblom: I know, but I think I would, in a way, spoil all the hard work I've put into researching earlier areas if I would now all of a sudden improvise a few remarks. I promise you it will come. That will not be such a long-term process.

Final Discussion

Transcript

Wallerstein: I've had the feeling, listening to the last two days of discussion, that they mirror very well the state of world social science. In my view world, social science is in the middle of an intellectual storm. We have gathered here as a group of people who feel uneasy about where things have been going. I also have the sense that people are unsure of the full dimensions of the storm. And that as we get more sure of it we, ourselves, get more frightened of it.

Professor Arnason yesterday said that we are observing the deconstruction of society.[1] He meant of a concept of society. We may also be observing the deconstruction of the real thing. Actually I think we are observing the deconstruction of sociology. By 'sociology' I don't now give it a narrow meaning, but a meaning of history plus all of the social sciences. I think they were slowly constructed for 120 or 130 years and they are collectively in a crisis now in which they are being not so slowly deconstructed and then, hopefully, reconstructed. We are faced with a nineteenth-century

[1] See p. 176 above.

Transcript (✉)
Berlin, Germany
e-mail: Artur.Bogner@web.de

A. Bogner, S. Mennell (eds.), *Civilisations, Civilising Processes and Modernity – A Debate*, Palgrave Studies on Norbert Elias,
https://doi.org/10.1007/978-3-030-80379-7_16

legacy which, we feel, doesn't work anymore. There's not only the concept of society which, I think, was central to the nineteenth-century legacy which I surely do not think works for all the reasons and more that were suggested yesterday. I also think that the nineteenth-century concept that the key methodological strain in social science was between a nomothetic and an idiographic point of view, which has informed all our works and, therefore, everybody has more or less taken sides in that great debate, also doesn't work. That not only is neither the nomothetic nor the idiographic point of view correct but neither is even relevant. And finally, I think—and this has been adverted to much more directly here—that we are, as someone said yesterday, unhappy or uneasy about the concept of progress while manifestly not being willing to abandon it nonetheless. Uneasy, but holding on to it in many ways.

So I thought I would start by thinking of the places in which we agree. 'We', now, is the group in this room. And then perhaps underlining where I think we are in some disagreement or uncertainty or debate, even though it hasn't always come out as clearly as it might.

I think that what unites the people in this room—of course it is not accidental that it unites them because they were, in some sense, chosen on that basis—is that they believe that there are such things as long-term processes and that these long-term processes are in some sense worth analysing. That they are, indeed, possibly the most important thing to analyse—I don't know whether everyone will go that far—and that the analysis should, in some sense, be a processual analysis. Having said that, that's a lot, because that's already a minority viewpoint. That is, holding that viewpoint puts us in a very small, if growing, but very small minority in world social science today. Nonetheless, I think it's correct and I think all the rest of you think it's correct. And I do think that that is much the way of the future in social science. That is to say, whatever comes out of the intellectual storm in which I think we are, I think that will be part of the premise of future work. That isn't very much, when all is said and done, however much it seems when one is fighting with the majority of one's colleagues! And beyond that I'm not sure how much agreement there is. And so I tried to outline at least four areas in which I'm not sure there's too much agreement.

One which we have not spoken of explicitly, but we have referred to implicitly and which is a favourite theme of discussion for me, is the unit of analysis. Insofar as we are making comparative statements, which we spend most of our energy making, one has to compare something. And one has to decide what are comparable units. And it is not all that easy to decide what the comparable units are, especially if one does not like the traditional category called 'society'. And it is not so easy to just say, 'well they are holistic units' because in fact, if everything fitted together analytically so well, we would not have the problems we do. But in fact, if you start defining your unit of analysis in one way or another, the fit is not perfect with other variables. And that is indeed the subject of analysis. So, let me point out to you the three different kinds of bases of units of analysis that have, in fact, been referred to in the course of our discussion:

(a) There is a political basis, a unit of analysis that has been called a 'self-governing political unit'. There are possibly all kinds of operational problems of defining what a self-governing political unit is, but, assuming we could resolve that, we could in fact compare at different points in historical time, different kinds of self-governing political units and make whatever statements we wanted to about long-term trends or about the non-existence of long-term trends, and so on. That has many problems and, in fact from my point of view, that is what we have been doing for about 130 or 140 years. That is to say, as suggested yesterday, 'society' has meant—in fact *only* meant—the nation state, which has worked for the modern world. And then we have tried to find other comparable—not comparable, just arbitrary—political units further back and we have compared them. It is apples and pears—that is perhaps the classic expression. But, it seems to me, it also misses exactly the point of seeing whether there are not significant differences between those entities held together by overall political structures and those held together in some other way.

(b) Then there is another kind of basis. And that is a rather traditional one, which is a cultural one. And we have heard such phrases as 'the heritage of aspiration' and 'common religious past'. Now that is a perfectly legitimate possibility. It faces ever-greater empirical problems of defining what is a 'common …': where does one locate a 'common heritage of aspiration' and how does one know that it is there? A 'common religious past' is perhaps a little easier, although it runs into lots of problems of what gets included in or out of these various things. But it also runs us into the enormous problem that most of the units we are in fact familiar with are situses of multiple cultural heritages. And indeed, one of the major political issues of most of the units we are familiar with, is how these multiple heritages fit together.

(c) So I myself have chosen the third possible unit of analysis which is the one based on the division of labour, which is also classic. In that case I won't repeat my analysis.

I underline to you that there are three modes and it is rather important which we choose because it leads to defining quite different groups, as a consequence of which we define quite different trends. We will see the world—it is not a matter of indifference—if we do our analysis based on one or another kind of unit of analysis. And it is a sort of preliminary question that one has to settle in one's own mind. I don't think we are going to collectively settle it, but I point that out as an area of important latent discord which results in non-congruent analyses. Secondly, if, as I suggested, the nomothetic–idiographic distinction does not make sense, what is the alternative? Well, the alternative is a middle path. At one point in one article I called it 'the war on two fronts' in which you fight simultaneously against nomothetic and idiographic presumptions And that ends you up inevitably with some kind of entity to which you have to give some kind of conceptual reality. I call it an historical social system within which one seeks to state the principles of operation (which is the nomothetic tension), but which is sufficiently distinctive from other historical social systems (that is the idiographic tension) that one has to be prudent about the generalisations that transcend the historical social systems. Anyway, that is where I come out. But, if you come out there, and I think a lot of you have, in fact, come out there implicitly too, then you run into a problem that is rather crucial again, on which I don't hear much accord here. And that is periodisation of human history. It is not a minor problem. For example, yesterday, Bill McNeill was making some statements which made a great deal of distinction between simple and non-simple systems.[2] And Professor Elias jumped and said, 'No, no. I don't accept that. It is much more of a continuity. You are making it a discontinuity and I think of it more as a continuity.' Then, at another point, however, someone else was making a distinction with the breaking point at the Industrial Revolution and Bill McNeill jumped in and said, 'No, no. It is much more of a continuity. You are making it too much of a discontinuity and I object.'

[2] See p. 120 ff. above.

So we have some choices. We can see it all as one kind of blur, a slow kind of melting on a single continuum. And, indeed, in many ways, that is the logic of the classic nomothetic position. When you make universal statements about how human beings act in society, the assumption is that the historical reality underlying it, from one end of history to the other end of history, varies—along the continuum, to be sure, but it is the same continuum. And therefore, you use the same tools of analysis and so forth. I, myself, as I said, am very sceptical of that. I think many of you are, but, then again, you are going to have to make some decisions about breakpoints. There are classical breakpoints bequeathed to us, which we are, in various ways, uncomfortable with. I have made my decisions, again, I point out, in the sense that I made my breakpoint in terms of a criterion which you may not accept, which was in terms of the relationship of different types of social structures to each other: in the pre-8000 BC period in which I thought there were only mini-systems; in the 8000 BC–AD 1500 period in which I thought there were three kinds of social systems, but with the world-empire as the strong form; and the more or less AD 1500 plus period in which the world-economy becomes the strong form and then the unique form and I would argue for those as significant breaks. It follows that others are insignificant breaks. It follows for me that the Industrial Revolution, so-called, is a minor blip on a continuum of the modern world-system, which has been exaggerated out of all proportion. But that means that the break in 1400–1500–1600—it is hard to date it exactly—is much greater for me than it is for many other people. And I am not going to argue my case again. I point out to you that the decisions on the breakpoints in our periodisation have enormous consequences for the mode of analysis, and certainly for the final outcome of the analysis.

Then there is an issue which we have been talking about an awful lot, which is a sort of the theory of knowledge. Now I may have stated my position yesterday more strongly than I truly feel it when we discussed plausibility. First of all I want to say I don't think 'plausible' is a bad word. I use it all the time myself and I indeed think that analyses, to be correct, must be plausible: I was merely trying to suggest that plausibility is insufficient unless it is tied to a considerable empirical base. And here I bring a worry which I have had now a very long time about the way in which people conduct comparative analyses. I have the feeling, without pointing a finger at anyone—all present company excluded!—that most comparative analysis between A and B consists of taking a microscope to A, which one knows very well, and comparing it with assumed and vaguely acquired knowledge about B. And therefore the assumed and vaguely acquired knowledge about B is at a level of generalisation. First of all, it is probably less accurate than your knowledge about A. But quite aside from that, it is at a level of generalisation which is quite different from your knowledge about A, and so you observe differences which are *not* differences, you miss similarities which *are* there, etc. Now, I was suggesting that we run a big risk in analyses that compare simple societies—my mini-systems or what you will—with world-empires, with the modern world-system, etc., simply because, not intrinsically but extrinsically, our level of knowledge is and will inevitably continue to be at such different stages of possible visibility that we should be very prudent about the statements that would compare what we can possibly know very well with those kinds of systems about which most of our knowledge, or a heavy proportion of it, is speculative. I have nothing against intelligent speculation, under whatever name it goes. But I would feel more comfortable if someone not only told me that it is plausible that 'if X then Y' but also told me simultaneously, and with some tone that might convince me or some logic that might convince me, that 'if X and only if X, then Y'. That is to say, ruling out alternative implausibility, which is seldom done. It seemed to me some of the objections that were being made yesterday in our discussion about fire [were in effect] 'Yes, it is plausible but are not alternative scenarios also plausible? And how are you going to decide plausibilities?' In any case I was calling for prudence there.

And then we come to the question of truth. I don't think we can escape it. I pose the question, if we are not talking about teleology and truth, then in what sense can we make statements that something is progress over something else, which we are constantly making? I think teleology, in order to defend oneself against the charge—nothing nastier in the modern world than to be called a teleologist. This has been a dirty word for about 500 years at least. I mean, after all, it was those religious types in the Middle Ages who were the teleologists and we, modern men, have all gotten away from this. So no one admits to being a teleologist. At least I have never heard a scholar get up and raise his hand and say 'I am a teleologist'. Nonetheless, what we do in order to defend ourselves is to give a rather narrow definition of what teleology involves. And then say we don't fit the very narrow definition, which is usually the case. But we may fit a slightly more liberal definition. I mean, I think, Professor Elias's self-definition of his activity is one that, in fact, most of us use, coming nearer to an appreciation of empirical reality. The whole idea of approaching, of getting closer—well, as you get closer and closer, presumably there is a limit there. Now, of course you can say 'We will never actually reach the limit. There will always be a distance between our perception of reality and what reality is, but we're getting closer.' And, as all of you know, as you get closer, it has sort of asymptotic qualities. You can divide the distance by 50 per cent between where you are in the end and it is true that theoretically there is always space there, but it gets pretty hard to see. So I am not sure that, by using such a statement, we are not really saying that there is something towards which we are approaching, which is a statement of truth. Now, it seems to me, there is only one way out of it. And that is to take a somewhat more relativistic view of the nature of scientific activity, suggesting that the objective is not, in fact, the attainment of either law-like statements or closer conceptual reproduction of empirical reality, but to say that what we're trying to be adequate to is meaningful interpretations in terms of the problems of the historical social system in which we live. However, if you do that, then you are face to face with the *cui bono* question. I don't see how it's escapable. In which case I don't see how we can talk about truth without talking about values. And, therefore, it calls for a radical rejection of the separation which has been dear to social science in many ways. That, then, gets me to 'progress'.

Fourth question on which I think there is not much agreement. Yesterday we talked about the Australian aborigines.[3] I know nothing about the Australian aborigines. I have never seen them in person. I have read relatively little: I've read Durkheim on *The Elementary Forms of the Religious Life*.[4] I've read a few articles here and there. I have some vague notion of what kind of social system they have, but nothing serious. There was some debate about the Australian aborigines yesterday which took the following form. In effect, are they civilised? And we were first told, 'Look, they never got beyond the hunting and gathering stage'. And someone else said, 'Yes, but they have a very complex system A, B and C. It's adaptive so it's in a civilising project, process, etc.' Yes, but somehow nonetheless, I had the feeling that at the end of discussion everybody implicitly agreed that one had to explain, somewhat apologetically, that they had not moved forward to sedentary agriculture. Now, in an argument I'm going to find a little difficult to reproduce because I don't remember its details

[3] See pp. 237–42 above.

[4] Émile Durkheim, *The Elementary Forms of the Religious Life* (London: George Allen & Unwin, 1915).—eds.

so well, Marvin Harris, it seems to me, argued several years ago that in general on this planet Earth, the move from hunting and gathering to settled agriculture represented a significant human regression.[5] I can recall two ways in which he argued that it was significant regression: that the total amount of time that people spent in productive activity in their life, that is, in labour, in a system of settled agriculture was significantly greater than in hunting and gathering; and that the average diet of the average person was, from any nutritional point of view, much poorer in a system of settled agriculture than in a system of hunting and gathering. Therefore, it represented regression and he then explained how come people regressed. Because there was this independent variable—I forgot how he accounted for it—of population expansion. There were just too many people around to permit a system of hunting and gathering. They were forced into a system of settled agriculture. Now, I'm not going to defend Marvin Harris's point of view. I suggest it's a possible point of view. If that point of view is correct, then the Australian aborigines had the good fortune to be located in a particular ecology, because of their isolation or whatever, to be lucky enough to escape the forced transition to settled agriculture. All the rest of the world went down this path to perdition. It at least opens the last question of whether we should presume, now, obviously, as a consequence of staying at the level of hunting and gathering they did not develop writing systems. They did not develop what we think of as high culture. They certainly did not develop what we think of as modern machinery and technology. And you are perfectly free to say that more than compensates for the poorer diet or the longer working hours, but we are now—I point out to you—in a metaphysical argument about values. Is it worth it? Does this outbalance that? I certainly have argued and would be willing to defend, but not at this moment because it would use up all my time, that the transition from feudalism to capitalism in the Western world did not represent significant human progress and that, in fact, it represented regression. I'm willing to argue that. I'm not willing to

[5] The idea of hunter-gatherer societies having enjoyed 'affluence' compared with their agrarian successors became the orthodoxy among anthropologists in the 1960s; see Richard B. Lee and Irven DeVore (eds), *Man the Hunter* (New York: Aldine 1968) and Marshall Sahlins's 1966 essay 'The original affluent society', reprinted as chapter 1 in his book *Stone Age Economics* (Chicago: Aldine, 1972), but the idea is much older. See David Kaplan, 'The Darker Side of the "Original Affluent Society"', *Journal of Anthropological Research*, 56: 3 (2000), pp. 301–24.

defend the [Marvin Harris] position, partly out of ignorance and partly because I feel there must be something wrong with that, but maybe …

McNeill: Having left the Garden of Eden was a mistake. That's been known for a long time.

Wallerstein: Yes, I appreciate that. And that can be the conclusion that you draw, that things get worse and worse instead of better and better. I want to argue a very different position. I want to argue that things have gotten better, have gotten worse, that progress is possible but not inevitable, that progress has been made but regression has been made. Therefore, we have the four real historical choices. I remind you of what I said. But then I think we have to define what our criteria will be. One criterion is self-control. The more self-control the more progress. And I'm comfortable with that, but only up to a point because I want to know, again, *cui bono*, and collective self-control as well as individual self-control, in which case what collectivity? What is the relationship to egalitarianism? Is egalitarianism itself the symbol of perfect self-control, since everyone denies himself the possibility of getting more than someone else? I don't know. But again, I'm not sure that our discussion of science is very far from our discussion of values. I conclude by suggesting that we have a lot of unfinished work. Norbert Elias yesterday said 'You think I am joking, but I am in the middle of my work'. I suggest to him that he was not only not joking, but he exaggerated. You are not in the middle of your work, you're at the very beginning of your work, and we are all at the very beginning of our work … perhaps even further back than you, sir, but nonetheless we are all at the beginning. And one of the problems is that we're going to have to redo a lot of the work that's already been done in the last 120 years. I don't say throw it all into the incinerator. I by no means feel that way. But we have been led down a lot of false paths, not only false theoretical paths, but I could argue that a lot of the accumulated empirical data is false and deceptive data, that we can't even rely on just building on all that. For example, the statistics that sit in thick volumes on our library shelves that have been acquired with painful energy over 100–150 years. Much of it, I think, has to be reinterpreted and put in the context of very new statistics in order to be useful. So I have the sense that we are not at scratch but we are not too far from it. We should not sit and think that we're building on the shoulders of giants, to use the phrase that Merton likes so much,[6] that we can simply benefit from 150 years of social science and 500 years of general scientific thought in biology and so forth and go on from there. At least I don't think we can simply go on from there, I think we have to rethink our premises and, once we've rethought our premises, we have to look at the accumulated data and use it most judiciously. And then we have to start building new sets of data from scratch. And that's a lot of work. And we should approach it both with humility and, I suppose, with courage or we won't make it.

Mennell: I have some remarks on the connection, differences and similarities between Professor Wallerstein and Professor Elias, and therefore this seems an appropriate moment to jump in, before Professor McNeill speaks because I think my remarks bear less clearly on his work—with which, I have to admit, I'm rather less familiar.

Can I take my departure from what Professor Wallerstein was saying at the beginning of his remarks about the problem of units?—the problem of units for comparison and also the similar problem of periodicity, drawing boundaries between periods as well. Now, I think the similarities and differences might possibly be a little clearer if we remember that Norbert Elias's work, although centred no doubt on his most famous book, *On the Process of Civilisation*, is really of more general significance, in that he has tried to put forward what he sometimes calls a sociogenetic and sometimes a figurational sociology. That approach plays down the problem of conceptual boundaries of periods and units, although he emphasises the survival unit (or the 'attack and defence unit', as he sometimes called it), as a basic element. I think the reason that he plays it down is that the central concept, which somehow doesn't seem to have been much discussed although it's there in the back of all our minds, is the concept of 'interdependence'. And this is very much a key element in Professor Wallerstein's work. You might say that the notion of interdependence is a sociological commonplace, that the idea of specialised roles, reciprocal roles, interdependence in that sense is so commonplace as to be banal. But I think that one thing which we have in common in this room is a somewhat *more radical* use of the notion of interdependence. Perhaps Professor Wallerstein in his work likes more usually to say 'dependence', but nevertheless, the

[6]Wallerstein is alluding to Robert K. Merton, *On the Shoulders of Giants* (New York: Free Press, 1965), but Merton was in turn alluding to Isaac Newton's remark in a 1675 letter to Robert Hook that 'If I have seen further it is by standing on the shoulders of Giants', and the image can be traced back still further, into the Middle Ages.—eds.

idea of reciprocal interdependence is implicit in it, as it is in Elias's work. To speak of 'interdependence' doesn't mean that the balances of power within a relationship are equal. Usually not. And in fact, I think that both Professor Wallerstein and those who have been strongly influenced by him, and those of us who have been strongly influenced by Professor Elias, both make more dynamic use of the idea of interdependence, focusing on ideas like chains of interdependence and webs of interdependence and their growth and their spiralling and complexity—and sometimes shortening and simplification. How I think this relates to some of the other problems that Professor Wallerstein has just raised is that out of the notion of interdependence and what Professor Elias calls 'interweaving' (*Verflechtung*) comes the idea of 'compelling trends'. He says that out of the interweaving of numerous people's plans and intentions come processes which are unplanned and unintended by anybody—which have acquired, if you like, a dynamic of their own—that the process is 'blind' and unplanned, and yet is 'directed'. It is 'directed' not in the sense of being consciously planned but in the sense of having a directional quality to it. And I think that idea is common to both Professor Wallerstein and Professor Elias.

Where there might be a difference, tying up with what Professor Wallerstein has just been saying about values, is that in Elias's work this is also intimately related to the whole problem of involvement and detachment.[7] It relates too to what Godfried van Benthem van den Bergh has called 'the attribution of blame as a means of orientation'.[8] If one understands the unplanned nature of processes that arise out of the interweaving of people's plans and intentions, then the problem of blame attribution, which plays such an important part in politics, is diminished. The interpretative function is more powerful than the application of one's own values to what one is interpreting. In other words, the development from a position of relative involvement to a position of relative detachment is very much bound up with the process of development itself. And then, out of that, I wonder whether Professor Wallerstein is in a sense more involved in

[7] Norbert Elias, 'Problems of involvement and detachment' (1956), in *Involvement and Detachment* (Dublin: UCD Press, 2007 [Collected Works, vol. 8]), pp. 68–104.

[8] Godfried van Benthem van den Bergh, 'De schuldvraag als oriëntatiemittel', *De Gids*, 141: 9–10 (1978), pp. 638–60; shortened version in English, 'The improvement of human means of orientation: toward synthesis in the social sciences', in Raymond Apthorpe and Andras Krahl (eds), *Development Studies: Critique and Renewal* (Leiden: E. J. Brill, 1986).

blame attribution than perhaps Professor Elias is. But I mean he would also argue that one has to make explicit one's standpoint.

And that brings me up to a final point. It's this problem of concepts. I think that one element which I found most difficult to understand in Professor Elias's work when I first encountered it was that there is a constant process of developing concepts in relation to the social processes of development themselves. Not being a philosopher at all myself, it took me a long time to work out what the philosophical premises perhaps are. And that's complicated by the fact that Professor Elias is actually rather antiphilosophical and feels that it's a great mistake to get too bound up with traditional philosophy. But putting it very crudely, I think that his work does involve a kind of radical rejection of the whole Kantian epistemological heritage and a recognition that Hegel was at least on to something—even though Elias is very far from being a Hegelian, again in any ordinary sense. What it does involve is a rejection of the value of conceptual analysis for its own sake, which is the dominant strand in theoretical sociology, in Britain at least, and I think in many other parts of the world—the playing around with concepts unrelated to the world out there, to the material that you are really working with.

The gist of my remarks is that there do seem to be considerable points of contact between Professor Wallerstein's position and Professor Elias's. I hope what I have said has illustrated some of the connections and the bases of difference.

Wallerstein: Actually yes. I don't disagree with any of the three points, though I put each just slightly differently. The interdependence—you felt uncomfortable using that word because you said it was a sociological banality—and indeed it is! I'm not sure that there'd be any social scientist who wouldn't say that groups of people are interdependent. So that isn't what distinguishes our point of view. I mean, Talcott Parsons would be perfectly happy to suggest that people [are interdependent]. ... Yes, I realise that, but I think it's a question of how you get to the interdependence. And there I think you're right. We're on the same side and against many other people. That is to say, the issue—and it's been an important one in social science—is whether a society is an agglomeration of individuals that somehow come together in some process, or whether in some sense the society is creating the individual, which is what I think you mean by interdependence. And, in that sense, I think there is an important similarity.

On the business of involvement and detachment, if I may be a bit detached, I think it's with what temporality you take this. I would agree that short-run involvement is not usually intellectually productive. It may be productive in many other things but, in fact, it creates many kinds of blinders. But I'm not sure I know how anyone can escape longer-run involvement. And the issue, therefore, is not involvement versus detachment, but at what length, time period, you want to be involved. Are we involved in terms of analyses of ten years? Or 500? Or 5000? And I am not sure that Norbert Elias is not very involved. I saw that yesterday when he reacted to certain kinds of examples and said, 'You can't be serious', meaning, I thought that that was an involvement coming out, not a detachment.

So, the question of blame attribution. Yes, I think one will get much further in understanding what is going on in a given country in the last ten or 15 years if one steers away from analyses which involve suggesting that there are good and bad individuals at the head of this or that institution. And then one utilises more analytic long-term processes which account for the fact that X or Y occurred. However, if I'm going to make some kind of decision as to whether the shift from feudalism to capitalism was or was not in some sense human progress, I'm not sure how I can do this without something that someone else would call 'blame attribution'. In fact, what is an interpretation other than that?

Finally, on concepts I absolutely agree. There I think we are absolutely on the same side. I think that concepts are only heuristically useful and that they constantly evolve under the process of evolution, but I do suggest therefore that that gets us to our periodisation issue. You see, I don't think you can understand why I started out by saying that we're in an intellectual storm. I think we are in an intellectual storm because I think the social structure that we live in has evolved in a certain way to a certain point which creates that intellectual storm, which causes not a minor but a major revision in our conceptual apparatus. There are minor revisions going on all the time. Every day and in every way. I remember, in fact, when Norbert Elias came to Binghamton some five or six years ago … I told him some of the activities of my centre, one of which as to do with trying to analyse the changing conceptual apparatus that people have used to analyse social [life]. 'Ah', he said, 'Conceptology. Very good'. I haven't stolen the word entirely yet, but, if I ever do, I will give you a footnote.

Evers: I think Professor Wallerstein pointed out at the beginning quite rightly that one of the main points that brings us here together is the study of long-term processes. And then immediately after that he embarks on a sort of rather encyclopaedic 'enumerations' of major research issues that have been discussed here and seem to be also of major concern to him. If you look at these issues, four of them—the first three, I think—arise very directly out of his 'conceptology', as it's now called, by first of all stressing the unit of analysis. Of course there are different world-systems or different systems as he outlines it. Then you get out of these two problems of definition. The first one is a temporal one: when does system A stop and system B start?—that's the problem of periodisation. After you have defined the unit, the system, and defined when it starts and when it ends, then you can do a comparative study: comparing these various systems and seeing how they are different…

After these three major problems, we are treated to a short discussion of the problem of progress and regression. What seems to be missing here, and I sort of want to stimulate this point and really see what Professor Wallerstein has to offer on this count, is how does one system change into another? I think this seems to be the crucial question. Somehow there seems to be a sort of demonic quality in the concept of 'system'. Once you get into the game of using 'system' as a major concept, you get trapped within the system. You can't get out of it. You find it difficult, for instance, to argue along the lines I suspect Professor Elias would argue. It would be difficult to argue in terms of sociogenetics, of defining the long-term processes in history as maybe a stochastic process, out of which different configurations evolve that had not been planned. I think this is a point of view that's also now quite prevalent in physics and in natural sciences which we normally associate with 'systems'. Now they have evolved beyond the systems perspective. So I wonder whether you could start again for two minutes where you stopped, namely with your fourth point, because I think that is really the meeting point between the three speakers that are lined up in front of us.

Wallerstein: Well, in fact, I don't think of myself as a systems analyst. My back arches when people call me that. I carefully use the word 'historical system', and not 'system', to indicate my strong belief that systems are not closed, but, in fact are by organic analogy, self-destructive over time. … Freud said that all life is one long detour towards death. In that sense, all systems operate in that way. Or that's my translation and my version of Hegel. That is to say, what contradictions involve are short-run solutions which create long-run problems. When the long-run problems cumulate to a certain point, the system disintegrates. In that sense, as I tried to say two days ago, I'm very deterministic. While you're in the system, it is demonic. You can't quite pull out of it. On the other hand, the system begins to crack at certain points. These are ... it only cracks at one point, right? But it's a long crack, at which point ... you can only say two things: 1. it's perfectly clear that the system will collapse over a period of time; 2. the range of human choice is increased. And yes, I would be happy to use the word 'unplanned'. That is to say, it is perfectly unpredictable, on the basis of our present knowledge, where a future transition would lead us. I sometimes use the phrase, 'The realm of imagination about social constructs is greater than we allow', and new things have always been invented and I'm sure new things will continue to be invented. But I don't in that sense feel that I'm trapped. That is to say again, it all depends where you ... I mean, the damn Greeks laid out all these issues and we've been wrestling with them ever since, right? And we never escape them, so there's determinism and free will and good [people] of Aristotelian spirit. I come down in the middle on this as in everything else so that I say a bit of one and a bit of the other. And the way I've reconciled it is to say that the determinism quotient is very high as long as the historical system is functioning well. And that the free will quotient becomes reasonably high when it's in transitional crisis, which I think we are in now. And I have so argued. I therefore think, incidentally, that that relates to involvement and detachment. It becomes harder to be detached when you're in transitional crisis, even from more immediate things. Or detachment is an option which leads in certain directions and not in others. It also speaks to the degree to which you're ready, in the context of social science, to discuss what are standard philosophical issues. I mean, 150 years ago, 200 years ago we made a great effort to detach all the social sciences from philosophy where they were. After all, in the university previously there were no separate departments. We were all still in the faculty of philosophy. That's why we all have PhDs today. I mean that's the relic of that particular history. So we made this enormous effort to pull ourselves away, which is, I think, thoroughly justified in terms of the place in which the world found itself in the early nineteenth century. I'm not sure that we shouldn't make some sort of conscious effort to reintroduce directly into our discussions some of the problems which we have now relegated to a [verbal] form of appendix of the world university system, the departments of philosophy. And treat them less as pariahs and more as part of us or we as part of them.

Kilminster: I'd like to make a few remarks about Professor Wallerstein's position. For me, when I hear the word 'crisis', I reach for my gun. I'm tired of hearing the word 'crisis', whether we are in a crisis or a transition.

And throughout Professor Wallerstein's paper that he wrote, the ending remarks he's made, he has come back again and again to this idea that he's talking now about a storm. He's talking about, somehow the social sciences have to be deconstructed because maybe the world-system is in the same kind of state. And I'm always personally very suspicious when I hear about this sort of thing. And I don't know whether the storm that he sees could just be a squall. It may not be the storm of the depth that he thinks it is.

It's a pity that Foucault didn't come because I would have liked to have said it to him personally, but people influenced by Foucault, too, are talking about the possibility that the nineteenth-century *episteme* has now crumbled away and we're in a stage where discursive formations are reorganising themselves and this is why we're all talking past each other. I am personally very sceptical when I hear this kind of thing. I think my remarks really are not very well constructed because I'm trying to say too much all at once. But, that's one point: I'm very sceptical that we are in such a stage at the moment.

Now the other point I want to make, to cut through to the central point really, is about periodisation. Professor Wallerstein said, 'We have decisions to make about this. I have made mine', as though this choice was an easy one to make—that one simply made a choice that there were periods here, periods there, or whatever. It seems to me that the problem with this isn't so much the choice of conceptual framework, but rather that the structure of the reality should determine what could be the choice of framework. He makes it sound as though this were a kind of decision, an arbitrary decision, that he can take. Now when you look carefully, the criteria upon which he bases his choice, it seems to me, are firmly in the Marxist tradition. I mean, his analysis is entirely, it seems to me, about relations of production and contradictions of this kind. And it's that that determines the discontinuity that he sees. His stress on discontinuity is because he sees a shuffling of modes of production …

Elwert: Could it be that this observation you made about Wallerstein has something to do with the present state of [words]. Where some people argue, and I would join those, that there is in fact a culmination of several crises. And that the failure of social sciences to cope with this originated some new reflections upon the conventional …

Stauth: … it's related to the statement of Stephen Mennell. You should ask the question, 'Why is Elias's work becoming so important today?'. That means the stressing of interdependence is obviously related to a stress of autonomy and dependency. What I don't see, is any sort of a formal agreement here in this room. There is a total disagreement. If you rely on division of labours, you rely on a certain autonomy in development, you rely on a certain self-development which is not always under control of people. Now, if I speak of interdependency, and interdependency in the way as Elias puts it, then I have a totally different type of setting or a new vision of humanity, as he puts it, acting in its concrete environment. And I think, we should stress this disagreement on this level or at least stress it for further debate.

Wallerstein: Well, let me answer all those briefly. I think it's very wise to be suspicious when people talk of crises. I myself am quite suspicious. The word is extremely overused and is used with some regularity. It doesn't follow from that that the concept doesn't refer to something real. In fact, I have gone to some great lengths in one long essay to say that most crises that people talk about are minor blips and should not properly be called crises. But there is such a thing as crisis, and it has to do with my concept that historical systems have lives and the crisis comes when their life is coming to an end, and the problem is the transition. ... And that, then, has something to do with the structure of reality. And so, of course, periodisation has to do with the structure of reality. But, of course, the structure of reality that we see is a function of the glasses that we wear. I mean, unless you're going to revert to a kind of simple Lockean view of the world, I don't think how we can separate our conceptual glasses from reality. Our glasses are part of reality, etc. And I was suggesting that if our glasses, which we have been using for the last 150 years, are not functioning so well these days, it is because the reality has changed. And, if we wish to function somewhat autonomously, then we need to get some better pairs of glasses. That seems to me a reasonable position.

I confess, I don't understand Georg Stauth's point because, if I start with the division of labour, that is in order to define the zone which I look at. If I don't start with that, I have to start with something else. I can't start with interdependency because I've got to decide what it is that, whose interdependency I'm going to measure. And, whatever I start with, I come up with a different unit. Once I've got my unit I can be very interdependent. I bow to no man in the degree to which I think the analysis of social science has to be unidisciplinary. I, after all, am advocating the abolition of sociology, political science, economics, etc., as separate disciplines, whether only it has any structural meaning or any intellectual meaning whatsoever. I don't think for a moment that you can analyse political, economic, and cultural spheres as in some way separate from each other, prior to each other. They are, in one sense, a deeply interwoven seamless web for me, such that I would restructure the entire university system to reflect that. And it gets me into more trouble than just doing it in my own work. But I still have to decide what the unit is that I'm going to analyse. You can say from here to eternity that you don't want to do it. But when you don't want to do it, you do it *de facto*. In fact, you can say for 150 years, 'We don't want to do it'. The very word 'society' was chosen as a nice vague word. Who knows where a *Gesellschaft* is? But *de facto*, when people started to actually do any empirical work, then they all knew where a *Gesellschaft* was. And its boundaries stopped where the customs frontier was. And everybody knew that *de facto*. But nobody would admit it in the elementary textbook that they gave to undergraduate students to read. So we can't avoid the issue. I want to put the issue right up in front. I want to say, 'Listen, this is an issue. And let's talk about it. Let us debate what kind of unit is a sensible unit.' And again, I've opted for mine. Not here immediately. But you can't work unless you opt for one. And I have self-consciously opted for one. Others of you have less self-consciously opted for one. And I don't think we're going to get a debate, a sensible debate going, until we drag out of people's subconscious on to the level of consciousness what they're doing.

Elias: I agree with Professor Wallerstein that one can speak of the present situation of the social sciences and beyond as a storm. But, perhaps in contrast to him, I think we shall weather the storm.

Wallerstein: Oh, I do. I agree.

Elias: In fact, what we have done here gives me courage. I think this shows that the patient is very much alive. And, in fact, it seems to me very odd that Immanuel Wallerstein, who has done a pioneering effort in the social sciences, should be so sceptical and pessimistic with regard to the past. This is a show that we can go on. It shows even the direction in which we can go on. Of course, like all pioneering works, it has its flaws. They will be recognised in the course of time. And there will remain some core which we can call advance in the social sciences. There is such a thing as progress in the social sciences. And I think what he has done—and, if I am allowed to say it in the same breath, what Professor McNeill has done—are pioneering efforts of showing how one approaches the problem of long-term processes. One can no longer do it in the speculative way in which it was done in former days, but solid empirical work which can found the pattern. I have done my bit, too. So I think the furrow which we plough is a very good one, a very promising one. Where I do not agree is that there is so much difficulty about the unit, shall we say the unit of reference, with regard to which we do our investigations. I think myself that Professor Wallerstein has been unlucky that the term 'system' was just in the air and that he picked it up and used it for something which really has not very much to do with the system. Sociological system theories abound, so why he did not find a more suitable term than this term which exposes his work to grave misunderstandings, I do not at the moment know. However, the fact is that it is perfectly clear to anyone who has eyes to see what the units of investigations are. It is only through the academic language, to the academic tendency to draw up a barrier between what we actually see and our perception that this problem is open and not clear. It is perfectly clear that what historians investigate and what sociologists investigate when they speak of 'society' traditionally are state units. The unit of reference of historians is states or groups of states. I do not think that is the only unit of reference—the unit of reference could also be a tribe or a band of aborigines. The theoretical question really is what have all these units of reference all in common? Why just state societies? Why just tribal societies? What distinguishes this kind of society, this kind of grouping from others? I have made an attempt to formulate it. As you say very rightly, I am only at the beginning. I know that. I have made an attempt, a provisional attempt, to say what these various units, such as tribes or states or bands have in common. I myself, in need of a more general term, called them 'self-ruling groups'.[9] And this is a provisional attempt to say what they have in common. Self-ruling or survival groups whose members, through their language and/or other marks of personal and social identity, are more closely linked to each other than to members of other groups. They are groups with decision-making agencies, temporary or permanent, not subordinate to any other, even in matters of life and death. And especially not in matters believed to affect the group's survival. As a rule, group members are, through customs or laws and/or reciprocal sentiments, better protected against physical violence within such a group than without, and are likely to act either under pressure of other group members or through pressure of self-restraint in order to ensure the survival of the group even at the risk of their own lives. I must

[9] See Elias, *What is Sociology?* (Dublin: UCD Press, 2012 [Collected Works. vol. 5]), p. 134. He more commonly used the term 'survival unit'. His most extensive discussion of it is in *The Society of Individuals* (Dublin: UCD Press, 2010 [Collected Works, vol. 10]), *passim*.

ask you to believe that this is not a statement of an ideal. It is anything but my ideal, what I have described. It is simply an attempt to state in more general terms common characteristics of such units which, to this day, we see as the main survival or self-ruling units. If I were to write a history of humanity or of the past, I would try to show how there was a time when human groups dispersed over the whole earth, inhabiting all the niches there are. How their languages differentiated in various ways. And how they again and again get into conflicts with each other. At first, I am sure the spaces [between them] were wide, ... but when they multiplied then the spaces became closer and the various survival units and the various self-ruling units got in each other's way. And there the structure of permanent war-like conflicts between the various survival units emerged. It is for me one of the most astonishing facts how little attention social scientists have given to physical violence and physical security as determinants of the groupings of people. We see it all the time. We are constantly confronted with fears today: the Russians and the Americans constantly fear that they could be weaker than the other. And therefore they are involved in this double bind of a pre-war situation, which I hope will not come off. But, in any case, I think that social scientists will not do their duty if they don't pay attention to questions of life and death, of physical threat and the physical protection against threat. Of course, one of these protections is, for instance, your protection of food supply, of goods which one needs, without which one cannot survive. So there is not really necessarily the distinction between 'economic' and what one usually calls 'political'. And I can hardly any longer use this distinction because, in early state societies, the economic and the political side are all rolled into one. And one cannot distinguish them. So it is not possible to use our vocabulary, the vocabulary of a very differentiated society, in order to describe societies which are not equally differentiated. There are questions which Immanuel Wallerstein has raised and which are very important, which seem to me very important in my own mind.

Perhaps I should say, before I go on, that when you say I am involved, of course I am involved. But if you look carefully at what I have written, I even start by saying,

> One cannot say of a person's outlook in any absolute sense that it is detached or involved (or, if one prefers, 'irrational', 'objective' or 'subjective'). Only small babies, and among adults perhaps only insane people, become involved in whatever they experience with complete abandon to their feelings here

> and now; and again only the insane can remain totally unmoved by what goes on around them.[10]

One can only talk in changing balances. And that, of course, is one of the centrepieces of my own theoretical approach that I try to get away from polarities and replace them by changing balances. So we have to adjust our conceptual framework to the very fact which we observe: that we have to deal with balances and study these balances.

Now, that of course, I think, is necessary to say because as you said very widely the social sciences have struggled to emancipate themselves from philosophy. I am struggling very hard to emancipate the social sciences from political philosophies, from political beliefs. I think that is the crisis within which at the moment we are. I have, in a playful manner, written this little utopia in which I tried to describe a new division of labour between social scientists who are not allowed to take part in politics and politicians who, if they need advice about the structure of societies, should try to take the advice of social scientists who go about investigating how things really are.[11] Of course, that immediately involves me in the question, which you also raised, of epistemology. I think, if I live long enough, I will be able to put a good deal of this epistemological unrest ... to give us all a greater peace of mind with regard to the question of what one formerly called 'truth'. I cannot, in this context, say it very clearly, but I have already alluded to it. It really needs an answer to the question which traditional theories of knowledge never pose and never answer. That is the question 'What is knowledge?' Professor Goudsblom has given us a very lucid idea of the uniqueness of hominins in handling fire. But there are some other unique features. And among the unique features of human beings is the capacity for forming symbols. That goes very far and I have no time to expose it here. But I think that the question with which we are confronted in epistemology is, how far the human-made social symbols can approximate to reality and how far not. So it is necessary to have a

[10] Norbert Elias, *Involvement and Detachment* (Dublin: UCD Press, 2007 [Collected Works, vol. 8]), p. 68. Elias is quoting from an article first published, in English, in 1956. In the conference, he gave his own off-the-cuff back-translation from a German edition. We have substituted the English original.—eds.

[11] See Norbert Elias, 'Knowledge and power: Interview with Peter Ludes' (1984), in *Interviews and Autobiographical Reflections* (Dublin: UCD Press, 2013 [Collected Works, vol. 17]), pp. 235–46.—eds.

symbol theory,[12] a theory of social symbols in order to say clearly what this approximation is, and the odd thing, of course, is that the symbols in no way resemble reality. They are something which is not commensurate, which is not like reality. I always like to give the example when Galileo was able for the first time to get a mathematical formula for the laws of falling bodies. The mathematical formula was more reality congruent than any human symbol before. But one cannot say that this mathematical formula resembled reality. In fact, it becomes more unlike reality. And that is one of the difficulties we have. Our concepts, our conceptology, do not resemble reality. But they symbolise reality in a purely human and social way. This is part of our uniqueness, that we can make symbols, collectively, of reality. So I am not able here to say more clearly how the greater approximation to reality of symbols can be achieved, but it can be achieved and has been done. Again, it is only because a blockage of cobwebs of words of our traditional knowledge have been interposed between what is fairly clear and easy to recognise if one only brushes aside many of these traditional cobwebs.

Unit of analysis: Yes, you mentioned the problem of comparisons. Of course, much of my work has been comparative work. If I say that in 1680 the standard of table customs was this, in 1720 it was this, and in 1750 it was this, that is a comparison between different standards. And the ability to construct a process model is entirely based on comparisons. Such comparisons are perfectly possible, and it is also—I fully agree with you—the development of a better theory of comparative studies, in that respect, is very necessary. I have recently done such a thing with regard to time or, as the word 'time' is also burdened with too many philosophical associations, the English language fortunately provides us with a verbal form, which is much more appropriate to what we say, the activity of 'timing'.[13] Now, one can, in certain respects, sometimes get hold of a process, of the pattern of the process, by simply taking the timing, for instance, of simpler people and the timing in our society. That will come out sooner or later, I mean, so I do not talk of something which I have not more closely investigated. One can get a very clear picture of the development of timing if

[12] Norbert Elias, *The Symbol Theory* (Dublin: UCD Press, 2011 [Collected Works, vol. 13]).—eds.

[13] Norbert Elias, *An Essay on Time* (Dublin: UCD Press, 2007 [Collected Works, vol. 9]). The earliest version of this book had been published as a series of essays in Dutch in the journal *De Gids* in 1974–75.—eds.

one uses examples from simpler societies. I use, for instance, the watch of a priest for the new moon. That is a constant experience. Wherever you look in former days, people had no concept of a regularity of the heavenly bodies. They had to see that the new moon really appeared in order to do the cult, or whatever festivals they wanted. So you can see the priests watching for the new moon, and there is, in this case, a very nice example. When his family also sees the new moon, they all bow and say, 'Welcome, moon. Welcome to our village'.[14] It is a feeling that the moon comes as a guest to their village. And when the priest goes into another village, he looks at the sky and finds it strange. The level of synthesis at this stage is so low that it is clear that the sky of one's own village is not the same sky as that of another village. And so we can see the problem how the enormously high level of synthesis of our society has emerged and also the enormously close-knit bondage of time, from which we often suffer. How has it come about that we are surrounded by watches and clocks and time is always there? Auden says somewhere:

> Time watches from the shadow
> And coughs when you would kiss.[15]

It is a very good image. This constant disquiet because the time is [always] there ... because in simpler societies timing is patchy. They don't need a time grid over their whole life. So ... I hope that Professor McNeill will, in that respect, agree with me that by means of comparison one can develop a very clear picture of the pattern of changes and the direction of changes.

I will not say very much more, I have talked about the epistemological side and about comparative study. You will forgive me if I use this opportunity, as I may not have another opportunity to do that, to say a word of thanks to the staff of the Zentrum für interdisziplinäre Forschung. I think many of you have been helped by the staff of the ZiF and by Dr Sprenger, who has done so much to help in the organisation of this. And I hope you will agree that in your name, I can give them your thanks also. I need not at this moment say thanks to you because there we are not yet at an end, but I wanted to at least say this. And perhaps, if I shall not take as lively a

[14] Ibid., pp. 135–50.—eds.

[15] W. H. Auden, 'As I Walked Out One Evening', in *Another Time* (New York: Random House, 1940).—eds.

part as I would wish in the coming discussions, you will forgive me. It is the burden of old age.

McNeill: Well, I feel first I risk anti-climax in the most drastic and extreme sense. I also have had a sense of being privileged to be here and belonging to an intellectual tradition that sets me apart from most of the people around the table, a kind of outsider because, as an historian—even if a slightly eccentric historian—the familiarity or the readiness to embark upon general statement and systematic analysis of what it is I'm doing, what other people are doing, is much less. I was much struck in listening to Immanuel Wallerstein this morning[16]—he moves immediately to a very high level of generality and the four points he made are indeed points which I will touch upon in the remarks I prepared, but I do it much less systematically. It's an interesting example, a difference between acculturation to the sociologists' culture as against the historians' culture.

Well, first, I am pleased to have been here. Not just for the change of scene and the different intellectual ambience, which is perceptible but not as strong as it might have been or would have been a few years ago, as in crossing the ocean. I've learned a variety of interesting things: the existence of penance in Eastern churches in the fifth century and why the concept of 'civilisation' was unimportant for Durkheim and Weber. And how very complex in unexpected ways the adjustment to the use of fire was for human beings when these things were first made. So papers of this kind, the papers we heard yesterday, were indeed interesting and various and stimulating. But the real issue for me is how to understand or how to relate my own ideas to those represented by Immanuel Wallerstein and Norbert Elias. The first of these, Mr Wallerstein's ideas, are in a sense much easier for me to cope with because I have had now—we've figured out—eight years of encounter since I first reviewed his first volume, and I have met him on several different occasions and had time to think about his great book and how my views compare and how they differ from those he advances. My general sense of enthusiasm for his work rests upon the recognition that I make, at any rate, that the structure of core–periphery–semi-periphery—which, if he didn't devise it, he at least brought it to my attention—is a real and useful anatomy of what in my vocabulary had been 'metropolitan centre' and 'cultural slope'. He gives it a much greater precision and structure than anything that I had been aware of before. And I think for many points of view it is a convincing anatomy of that interaction

[16] See pp. 251–2 above.

between centre and the extremes of the different levels of what I formerly called 'cultural slope'. I do, nevertheless, have certain reservations. This will repeat things which I have certainly said before, but I hope you'll forgive me for. First, and I think quite importantly, the system seems to me much older than he has portrayed it: there was, I believe, a period in which China, South China, was the core and the Mediterranean was the periphery. And later this centre moved to the Mediterranean, and first the Italian cities, for something like three centuries, were a core and north-western Europe and other parts of Europe were a periphery. And only from, say, the seventeenth century rather than earlier can you locate this core in the area where Mr Wallerstein locates it, in north-western Europe. So I think that the world-system, the modern world-system, has a much older history. I think you had to date it from the eleventh century. And there are foreshadowings of it throughout the first millennium, the Christian era as well, as I decided to say in my paper on Friday. So that's the first reservation I feel about his book.

The other reservation is something Mr Evers really hinted at, too. It seems to me that the political–military impulses, principles of organisation are not as subservient to economic and market relationships as Mr Wallerstein seems to say. I think there is a much greater autonomy for armed leaders and followers than his book implies. And probably also of intellect—that is, the notion that our idea system is, so to speak, frozen into a world-system which is primarily an economic system or a division of labour system, seems to me implausible. I think there's much more scope for the free imagination and the choice which can run against the grain of economic interest. I mean, many people have preferred non-economic returns for their lives. And, to some degree it seems to me all professors are in that class. We don't try to maximise our income. At least I've never been conscious of trying to do that. In fact, quite the contrary. I feel I have been emancipated from the necessities of the market by the position I hold, getting paid a salary every month. I don't have to worry. I have more money than I need. And I do not, therefore, feel myself as subservient to the economic system of the world or the capitalist system, as I think Mr Wallerstein thinks I am. Now that may be self-delusion. I do understand that his followers have accused me of blindness. More generally, I would argue that human motives seem to me far more complex. There's a spectrum of motives, from hate to fear and love, that are only very thinly and imperfectly registered in the marketplace. And these motives do move human beings, and don't disappear when an economic system of long-distance trade and flows of capital investment move across the long distances that the modern capitalist world has made possible. And I feel there's a real limitation of your outlook. I mean this is another way of saying that you are not a Marxist, at least an ex-Marxist. And that Marx focused on a critical segment but not the whole segment of human behaviour.

The third point is something that I share with you is that you pay little attention to the ecological nexus in which human life exists. And these things do intervene in the market and economic world quite dramatically. Population growth, disease, changes in disease incidence, things as I've put forth in my *Plagues and Peoples* book,[17] it seems to me, ought to be part of your conceptual world and aren't. So that's the degree of limitation I feel in my understanding of [Wallerstein's] book. But let me return and say, I think it's a great book and a book which you should be proud of and that I think will become, will remain an important book in the future dialogue between historians, sociologists and others.

Far more problematic for me is the effort to grapple with Norbert Elias's ideas, largely because this is very recent. Until he sent me his books in anticipation of this conference, I had not read them. And so I have a much shorter exposure and experience, and, when you get old—perhaps you don't have to be old—it takes time, you have to sort of feel your way around things and think about them. So I have a far less firm and clear notion of my reaction, my understandings, and my reservations. But let me try anyway.

The first thing that I admire is the specificity—the cases, the down-to-earth, the real people and the real people's lives, and real evidence, the etiquette books—as against the effort to put the world into more abstract categories—and a level of argument [about] detailed everyday sort of tangible human experience. And this I think is a very, very important characteristic and one which has tended to isolate sociologists from historians. As we tend to get a lot of data, you tend to get a lot of ideas. Too much data with no ideas is a distress. Too many ideas with no data, or with the data so generalised that you can't say 'where do you see it? Where is it?', is at least for me indigestible. I run away from it.

The second thing is that, as I accept from Mr Wallerstein his notion of core–periphery–semi-periphery as a very helpful and persuasive and powerful idea, for a more formal analysis of what I'd previously simply, loosely,

[17] W. H. McNeill, *Plagues and Peoples* (Garden City, NY: Anchor, 1976).—eds.

not anatomised, so too I think is this notion of 'figuration' and its power to shape behaviour of relatively small specific groups. *The Court Society*, I thought, was a genuine virtuoso's work. And, of course, it helped to have Saint-Simon as your source.[18] But really, the sensitivity with which you describe the courtier's life is something that I had never come close to before. That sense of yours, 'this is how it really was', came home very strongly to me. And, thinking about this, I believe that human society, human life is very strongly affected by this sort of little clusters—not always very little—of people whose mutual interaction defines the behaviour of each and, to a very large degree, the consciousness and the personality. I remember when I was a graduate student—and then was drafted into the army. I thought I was an autonomous human being. And I brought my books with me, or at least a small selection of my books and thought that after hours I would read books and be myself. Well, within I should think about six weeks, I discovered that wasn't true. The first problem was after hours there was no place to read. There wasn't a light in the barracks. And the men around me were not reading books. If I insisted on reading a book, I was peculiar, and it took a tremendous effort to withdraw from the herd and I gave it up. Now, I was quite surprised at myself. I thought I was an autonomous human being and it turned out I wasn't. I became a soldier.[19] And I can also remember the time I was discharged from the army, five years later. And the first, I don't know, six weeks or so of just being myself, just a civilian—all of sudden I'd lost an identity. The things that defined my rank and my relationship to other people, which had been defined by the clothing you wore and by the marks on your uniforms, now were gone. And who in the world am I? That sense of nakedness—I will never forget that first few weeks: before I had been a something, now I am nothing. Or at least I'm not the same anyway. And it took quite a while to become a graduate student again. I made the transition successfully but this is a very vivid memory of that power in my own life of this sort of figuration. And you know what happens in a university, assimilation

[18] Saint-Simon, Louis de Rouvroy, Duc de, *Mémoires complets et authentiques du duc de Saint-Simon sur le siècle de Louis XIV et la Régence*, 40 vols (Paris: H.-L. Delloye, 1840-4). [*Mémoires*, ed. Gonzague Truc, 7 vols, Bibliothèque de la Pléiade (Paris: Hachette, 1953–61). Partial English translation by F. Arkwright, *Historical Memoirs*, 6 vols (London: S. Paul, 1915–18).]—eds.

[19] McNeill also referred to the lasting effect of his experience of military training at the beginning of his later book, *Keeping Together in Time: Dance and Drill in Human History* (Cambridge, MA: Harvard University Press, 1995), pp. 1–4.—eds.

to a particular department, a particular style of life. It's also the same that happens. And I am very glad to have met *The Court Society* and to have a much clearer and firmer idea of what [happens], and I think the term 'figuration' is also a very useful term to describe it.

Now that's the affirmative side, those things which I feel I have learned and profited from.

I do have doubts about the notion of progressive refinement of manners across many centuries. Greater suppression of impulse, greater self-control—I think rather it's a series of ups and downs. That there are societies, what I called 'simple societies' yesterday, where there's a very high level of self-control and a very highly specified code of manners. And without a written record it's really very difficult to penetrate. You saw what the anthropologists thought in the latter part of the nineteenth century. I don't think it was all a myth though I think there's a good deal of wishful thinking in their projection upon the Trobriand Islanders and the Kwakiutl and whatever else they were studying. Of all the things they did not find in their own societies, they tended to project it upon the people they were studying, to some degree. Margaret Mead in Samoa I guess is the clearest example of that.[20] Nevertheless, I think in such societies a very high level of personality definition, a highly codified code of custom did prevail.

Now what you often see in the civilised situation is, when people with different customs, with different codes collide with one another, the breakdown of such codes. So instead of civilisation being a restraint, it sometimes unleashes [barbarism?].[21] ... [A] beautiful example of this is where the [under the impact of Spanish conquest] the Amerindians' rulers lost pretty near all morals.[22] And the recovery from that was indeed a slow and gradual process. And in more modern times, the pattern of behaviour that started Toynbee off in his notion of plural civilisations, when he saw what the Turks were doing to the Armenians and the Greeks were doing to the Turks in the First World War. He concluded that what you had here is a breakdown of moral code, what he came to call 'amoral society', 'amoral civilisation'—breaking down, leaving its heirs with no moral

[20] McNeill was alluding to the studies of three famous anthropologists: Bronisław Malinowski on the Trobrianders, Franz Boas on the Kwakiutl, and Margaret Mead on Samoa.

[21] Several words are missing from the transcript about here, but the meaning is clear enough.—eds.

[22] The tape recording is unclear at this point, and we cannot be sure to whom McNeill attributed the observation or description of the 'beautiful example'.

restraints, or without the moral restraints which, in an older generation, they have had, now are gone through collision with the West. And the amoral conquerors of the Far East in the nineteenth century are another example. There are numerous examples of this—where moral codes break down through collision with an alien society's environment. And then—because amoralism is fundamentally extremely anxiety-creating, it seems difficult to live with—you get the reactions against such amoralism: reformations from time to time, of which the most recent is that of Khomeini, but by no means the only one. And I would even suggest that the rise of what we call 'higher religions', 'universal religions', between 500 BC and AD 700, is the great landmark of human reaction against amoralism which is the product of civilised mingling. It is the great institutional invention which allows strangers to encounter one another, live side by side and share enough of a moral universe to maintain, to allow the further elaboration of the impersonal relationships of civilised society, the bureaucratic relationships of the empire, and the market relationships of the economic systems, the world-systems. What Mr Hopkins was talking about yesterday was an example of this, it seems to me, a moral repugnance against an amoralism that had been characteristic of the Roman Empire, when many different peoples came together and they had no common moral code for a while, or a very slender one. What the Stoic philosophers were doing was the same thing, but their resonance, their public, was smaller than that which fell into the hands of the Christians. And the paper[23] we didn't discuss, but which I found an absolutely fascinating one, that by Elwert, 'Markets, venality and moral economy', is another example of this kind of thing. ... The German Reformation was the same thing—in other words, an example of the same, a convulsive reaction against what you call 'venality', the breakdown of older moral codes and finding a new basis for community, quasi-community. And this, I think, is the invention of these portable religions, religions which go anywhere, with a universal moral code. Really the moral codes of all these religions boil down to how to treat strangers, do unto others as you would be done by; it allows markets and bureaucracies to function with a modicum of efficiency in the presence of plural strangers, people who are otherwise strange to one another.

[23] For a revised version, in German, see Georg Elwert, 'Ausdehnung der Käuflichkeit und Einbettung der Wirtschaft: Markt und Moralökonomie', in: Klaus Heinemann (ed.), *Soziologie wirtschaftlichen Handelns* (*Kölner Zeitschrift für Soziologie und Sozialpsychologie, Sonderheft 28,* 1987), pp. 300–21.—eds.

And this, I think, is the sort of architrave of what I want to call 'civilisation', and without it I don't think our civilisations would have had the history they did. So, my first reservation is that I don't think civilisation consists of a sort of smooth path towards civility, towards self-restraint. Rather, in a certain sense, civilisation is antithetical to that. It breaks down moral codes, and then efforts to find new solutions and the processes within the particular constellations, figurations, are examples of this response to and creation of a tolerable moral universe in what otherwise would be unsatisfactory conflict and diversity. That then is my first dissent, non-commitment. I don't know that I am right but I do not feel convinced of the notion of civilisation as restraint, increasing restraints smoothly through time, if that's what you say. Now, maybe, I'm reading things into it I shouldn't.

My second is the emphasis you put upon the state, the centrality of the state as the important human community. You just reaffirmed this in the remarks you just made. It seems to me almost the converse of Wallerstein's centrality of what I would call the 'market' and you, I think, call 'capitalism'. But perhaps we mean almost the same thing by 'market' and 'capitalism'. I'm not really sure of that. And where I said I see links that run beyond the state that are also important, and linkages or formations that are of a sub-political level that seem to me even more important, where it seems to me your figurations figure very much. The figurations that are tangible, that are real, seem to me mostly sub-political and every so often one of these figurations makes an alliance with the police power of the state, as the Christian church did in the fourth century, with very important consequences. But the police power of the state, especially in older times, was a limited one, before computers and instant communication and card files ... And I don't think that police power is ever all-important There is the continuing autonomy of local groups, as I'm sure you don't deny. And this is where we probably do differ. The real integument of what I want to call 'civilisation', the shared value, the style of life—in most cases, it's not within a single political group at all. Once in a while most of Chinese civilisation was under one emperor, but even there, if you look closely, there always are fringes which run beyond. The Chinese overseas, the Chinese under barbarian rule, to the north—so, I think there's never been a time in Chinese history when all Chinese were under one emperor. At least I would be very sceptical if that were the case.

So these are my reactions to date to the two idea systems that flank me. I thought it would be only decent to conclude by saying, it seems to me there are weaknesses in my own system of ideas which have been only casually and lightly treated. One is—and this, I think, Immanuel Wallerstein is probably too polite to say—there is a sense in which, you know, I am an apologist for US hegemony and running dollar capitalism.

Wallerstein: I'm not too polite—I don't even know the words!

McNeill: My ideas, so to speak, march with the big battalions—the wielders of power, the victors are the people that I'm interested in and tend to sympathise with. And the victims tend to be, you know, the waste bin of history. And you can say this of me—there is some truth in this picture. It is a limitation of my sensibilities and of my personality, no doubt. I've been struck, looking at my fellow historians [that] it seems to me we do divide, pretty obviously, between those who sympathise with power-wielders and those who sympathise with their victims. Historians are either one or the other. And very seldom is an historian able to sort of be, both. I would like to think I am both, that I see the victim as well as the victor but it is quite true that in my books, if you read them, it is the survivors, the victors, the people who didn't go down in the plagues that I tend to talk about and tend to identify with in a sort of subliminal way. So, this is true: I am on the side I am on, insofar as these are sides. And you can justly accuse me of this sort of identity bias.

The [other] we did discuss after my paper: 'what is civilisation?' How real and tangible is it? Is it a figment of my imagination? Or not just mine, but the imagination of those who've talked in such terms. I do indeed recognise that that is a term of fragile obviousness. My best effort is to say, it's a style of life, which is recognisable, even if it's very difficult to articulate in a particular case what it is you recognise that binds together different behaviours and makes them coherent—well, 'coherent' is too strong—a style. As in art history there are the marginal cases where things slope off and are touched by but not fully shared. There are the provincial bad copies of the metropolitan high style, with the whole degree of transition from a perfect example of a style to something that no longer belongs to the style at all. And that kind of transition in art history, if you are familiar with it, it can be done with quite a good deal of precision and very little argument. The experts know what 'baroque' means and they can draw the lines and the limits and the intermediate cases. And the minglings and the boundaries are not hard to draw. I think that's fair to say. And it seems to me that in social styles, style of life, there's something similar, though it would be nice if you had something as unambiguous as the visual record. Actually, in my own thinking about these things, I do use art very strongly in my own teaching of world history and civilisational history, and in my thinking about it. If I ask myself, 'What am I really talking about?', this is the sort of subjective definition. It's a style of life which you recognise, and you have very great trouble verbalising just what it is you're recognising, a family of behaviours. Theoretically I would try to say it is the result of a communications net; working within the limits set by animal transport and sailing ships and writing; and, of course, as I said on Friday, especially the sacred texts which tend to take a canonical and central role in the acculturation of those who share in the particular civilisation in question. And, if this is correct, of course, the civilisational unit may indeed dissolve as forms of communication falter. And, if so, then the sort of transformation of human affairs that Mr Wallerstein is preoccupied by—the coming crisis, or the crisis of the world-system in his language—may also be, for someone with my vocabulary, a time of the disappearance of the autonomy or the recognisability of the plurality of civilisations.

The third problem of my idea system that I'm conscious of, which I did mention before, but I thought I should recapitulate it for thoroughness, is this problem of the relationship, feedback loops across levels of organisation: the cultural, ecological, physiological equilibria and their interaction which I am certain occurs. But how to describe it? How to put it into words? It not only puzzles me, it staggers me. I have not done it to my own satisfaction. And I don't really think anyone else has done either. But, as I said, it seems to me that one of the limitations of your system is that the ecological level is really not there—but it *is* there in human life. And if you're going to do a proper sociology or a proper history, somehow the cultural, semiological and the ecological—and I must say also the physiological, the thing that keeps our bodies alive, all have to be somehow related. And their permutations, combinations, are very complicated work. We have not yet deciphered it all. And it's just as well—we wouldn't have nearly as much fun in it, if we really knew all the answers.

I thank you, Mr Elias, for bringing me into this context, for the sort of more vivacious recognition of both the limits and the congruence of my thought processes with those of my distinguished sociological colleagues to become a little clearer to me.

Wallerstein: I'd like to say two words. First, very briefly to Professor Elias: if he interpreted me as being pessimistic, that is a misperception. I am not pessimistic about the …

Elias: I am glad.

Wallerstein: I am prudently optimistic. And secondly, what you say about polarities and balances I am in total agreement with, and I think I was trying to make the same point.

And so, for the critiques of McNeill, first of all, I would like to say that on the business of whether the world-system started in the sixteenth century or in the eleventh century, I am open to criticism and have often reflected since on alternative datings. You are not the only one to have made that point to me, and it has some plausibility. That does point up the whole issue of periodisation. On the ecological nexus I plead guilty regretfully. On the other I plead not guilty; I think your examples are poor ones. The fact that you have different motivations as an individual from maximising your profit speaks not at all to the issue as to whether the accumulation of capital is not the driving force within our present world-system. That it is the driving force is perfectly compatible with the creation of a system in which that is not the driving motivation of every individual within the system. And you pay a certain price for your particular motivation which, however, if I may say so, is functionally useful to other people. So similarly with the political and military impulses which are certainly there, but don't have a separate life of their own, I think. So that comes back. But I do note that the two talks successively have illustrated my point that choosing the basis of the unit of analysis does lead you somewhat differently, because we got a strong reaffirmation from Norbert Elias of the state as the basic unit of analysis and from McNeill—a bit more vaguely but rather clearly on some kind of culturally held together nexus as the …

McNeill: I perfectly agree with you. It's the fundamental commitment, how you go on, and the way the world shapes will depend upon it. It seems to me that your statement beginning like that was perfectly right. And the chronology is similar. Periodisation makes all the difference in the world: what you put together and what you separate.

Elias: I just want to correct one point which you made. I never said that simpler people have no self-control. Not only have I lived in Africa[24] but I am fully convinced that civilisation is a human universal. What I think, however, is that the *pattern* of self-control is different. And that one can show in a very graphic way in using timing for comparison. In simpler societies timing is patchy. It is confined to certain occasions. The moon comes, and then when the moon is there one has no longer any need for

[24] Elias served as Professor of Sociology at the University of Ghana in 1962–4.

it until another point-like occasion comes where one times again. The same is true with self-controls. In our societies self-controls have to be like timing—all-round. They never leave us. They are evenly spread. I mean, I could go into a further description but I just wanted to say, in this context, the direction of the process of civilisation is not from no self-control to self-control. It is from a specific pattern of self-control which is point-like, with many relatively freer outlets for uncontrol possibilities, to the state which we live in where the conscience formation is, as it were, all-round. And we can only rarely escape from the allround cover. I mean, that is roughly, very roughly the direction of the process as I see it.

McNeill: Well, my question would really be whether in these hypothetical simpler societies there are these periods of relative uncontrol. This is what I'm doubtful of. But I don't know enough about such societies to really be dogmatic.

Elias: The possibility, shall we say, for killing someone, in times in which the pacification through one way or the other was not yet as strong as it is today, was very much greater than it is with us. The possibility to use physical violence …

McNeill: Except in certain parts of the south side of Chicago where young blacks kill one another every day. We have our moments of breakdown, too.

Elias: I'm not sure whether you mean that there are some people in which it breaks down, which is the case. Or whether you say that we all have a breakdown. And that I do not believe.

McNeill: No, we don't all. But in a modern city, in the city of Chicago which I know best, there are subcultures—if I can use that phrase—that are armed and shoot one another.

Elias: Yes, that is perfectly true, perfectly true. But there are also …

Wallerstein: May I not ask you, in the twentieth century have we not had collective violence of a rather extraordinary level? I don't need to cite it for you. You've lived through it.

Elias: Yes … but I mean, don't let us forget, don't let us confuse life in a village where there is a relative unevenness of control among all people. That is to say the formality which we see in simpler society is really a security. It gives security against something which is very insecure. It assures you that, if you go to a friend, you have first to chew some nuts with him or her. And that is, as it were, a wall built against the insecurity of what they can do to you. So in our society you have a good many people where the restraints never really break down. And you have others where they break down. That is a great difference. I mean, in order to point to the concentration camps …

Wallerstein: Or just to the people who kill each other in Greece and Turkey, India and Pakistan, Iran, Iraq, etc., etc.

Elias: Well, let me put it still in another way. In regard to interstate relations …

Wallerstein: These are not only interstate.

Elias: … nothing has much changed.

Wallerstein: Yes.

Elias: In regard to interstate relations, of course, violence is used in as brutal a way as the Sioux Indians do it. I mean, that is part of the theory—that in interstate relations we are still in the late Middle Ages.

Wallerstein: A lot of that was within states, you know.

Elias: Within the state the pacification has gone very far. You can see it in simple things. For instance that women can go along the street without being in danger of rape. Two centuries ago even in European countries it was often very difficult for women to go along the street alone. It was really not done by a respectable woman. So in our time it is possible for women to do that. That's one of many examples.

McNeill: There are parts of the south side of Chicago where it isn't.

Dunning: Professor McNeill, are you familiar with the work of Gerald Suttles on these areas?

McNeill: No.

Dunning: Well, he puts forward a thesis, particularly in his second book, *The Social Construction of Communities*,[25] which is consistent with Norbert Elias's theory of the civilising process. That is to say, he argues that the state, the arms, the agencies of the state are unable to control those areas. I mean it's perfectly consistent.

McNeill: They have an internal police structure, of course, among their own members.

Dunning: Their own, yes. I mean, that's his argument.

Elias: May I just say in this respect I am always a little perturbed by the fact that in talking of these problems, the short-term perspective, our focus on our present distress, is so great that we cannot distance ourselves enough to see what the difference between a simpler society and us really is.

[25] Gerald D. Suttles, *The Social Construction of Communities* (Chicago: University of Chicago Press, 1972).—eds.

Wallerstein: But, you see, it seems to me it's partly a question of what you measure as violence. Now we have some fairly good data—although it has come under some question lately—to deal with some of these socially constructed ecological variables that Bill McNeill likes to deal with, which are called *The Decline of Population in Mexico and Central America in the 100 Years following European Conquest* by Borah et al. [26] The Borah (et al.) figures, which I agree are controversial but nonetheless they are there, argue a decline over an 80-year period from about 15 million population to about 1.5 million, partly through direct killing, partly through the spread of disease, etc. Now that was a direct consequence of a socially initiated action that was part of the construction of my modern world-system. That was 90 per cent of the population, and those figures, which are perhaps the most dramatic—although there are equally dramatic figures that involve not disease but actual killing by guns and other ways that come up to such high percentages—have to be put into the analysis of violence. It's collective violence. It's concentrated in time, but if we now precisely take the long-term viewpoint, and the large-scale viewpoint, then we have to sort of count the multiplicity. I agree that for certain subzones, the cities in the core countries, life is safer, although it's been breaking down in the last 20–30 years, but you could measure a direct increase in the safetiness [*sic*] of the individual's life in London from say 1600 to 1900 and I think you would probably come up with a constantly increasing level. But London has to be put within a world. And therefore, to count London alone may deceive us. We would have to count everything going on. And then we would have to say, having counted everything that goes on, has the overall level of safety, not for the people living in London, but for the people living in my whole historical system gone up or down? I don't have a firm view. I'm on McNeill's side on that. I reserve judgement. I'm not sure.

Elias: Well, I agree with you that one has to take the whole world. I mean, I was very much in agreement with Professor McNeill that, at one stage, one has to take the cities of Rome and the nomads as well. One has to see them together. But what happened in our society, at our stage, at

[26] We have been unable to trace a book under exactly this title, but Wallerstein appears to have been referring to one or more of the following: Sherburne F. Cook and Woodrow Borah, *Essays in Population History: Mexico and the Caribbean* (1971); Woodrow Borah and Sherburne F. Cook, *The Aboriginal Population of Central Mexico on the Eve of the Spanish Conquest* (1963); Woodrow Borah and Sherburne F. Cook, *The Indian Population of Central Mexico 1531–1610* (1960), all published Berkeley, CA: University of California Press.—eds.

the world stage, is the astonishing thing that, in our society, all human groups have coagulated into states. This is something completely novel: never before has it happened …

Wallerstein: Yes, that's right.

Elias: … that they have coagulated into states. And we have therefore a differential between all these struggles, which go on in the newly formed states for integration on a higher level, which creates a lot of conflict. But this is a new organisational level into which they all have been put, if tribal groups get together into a state level. And here I want to say, Professor McNeill, I do not see what you meant before, when you said there were states, but there were very many who were not living in states. If they did, I misunderstood you there, yes?

McNeill: I don't think I said that.

Elias: In former days. I mean, you made a remark that there were states but many who did live outside. What my thesis would be, let me say briefly, is that very few existing people do not live in survival units.

McNeill: That I perfectly agree. Your term 'survival units' solves the problem of state versus the tribe or the community which is described as stateless. They still have a political structure and they still have to attend to their own survival somehow. So I think 'survival units' is a very good roof for these different levels of political articulation without necessarily having specialised officers who are in charge of the army and tax collection system, which we use as liminal of the state.

Elias: If I may say, that is the problem which we should investigate. We should investigate whether this hypothesis, this theory, which I have put forward … But for that one has to read it. One cannot simply take the popular vulgarised version that I say 'More self-control'. I may have slipped sometimes—that may well be. But on the whole in the second volume, where I give a preview of an overall theory of civilisation, I have concentrated rather on what I have described now.

Index[1]

[1] Note: Page numbers followed by 'n' refers to notes

A. Bogner, S. Mennell (eds.), *Civilisations, Civilising Processes and Modernity – A Debate*, Palgrave Studies on Norbert Elias,
https://doi.org/10.1007/978-3-030-80379-7

L

M

N

The manufacturer's authorised representative in the EU is Springer Nature Customer Service Centre GmbH, Europaplatz 3, 69115 Heidelberg, Germany. If you have any concerns regarding our products, please contact ProductSafety@springernature.com

Printed and bound by CPI Group (UK) Ltd, Croydon, CR0 4YY
30/09/2026
02225326-0010